LIZARD MAN

The True Story of the Bishopville Monster

Lyle Blackburn

Anomalist Books
San Antonio * Charlottesville

An Original Publication of Anomalist Books
Lizard Man: The True Story of the Bishopville Monster

ISBN: 9781938398162

Cover artwork by Carlos Cabrera

Illustrations by Joseph Daniel Patterson

Maps by Lyle Blackburn

Photos courtesy of individual photographers as credited

Book design by Seale Studios

For more information about the author, visit: www.lyleblackburn.com

Or go to AnomalistBooks.com, or write to:
Anomalist Books, 5150 Broadway #108, San Antonio, TX 78209

For Sandy and Lyla.

With special thanks to my exceptional research partner, Cindy Lee.

Contents

This book contains actual accounts by real people based on personal interviews, newspaper archives, and police reports.

Foreword

I personally consider Lyle Blackburn's *The Beast of Boggy Creek* to be the finest cryptozoological title of 2012. It quite rightly received excellent reviews and justifiably thrust both Lyle and the Fouke Monster itself firmly into the limelight. Sometimes, however, when an author puts out such a great *first* book it begs an awkward question: how do you follow it with an equally good *second* book?

Of course, Lyle could have done the literary equivalent of what the Sex Pistols did with their classic 1977 album, *Never Mind the Bollocks*. That's to say, the band put out an amazing record and then split only months later. Their legacy: a still-stunning, stand-alone collection of anthems from a group that went out with a bang.

Fortunately for us, the readers and fans of all things monstrous, Lyle chose not to follow the "live fast, die young" approach of the Brit punks. Instead, just about as soon as he was done with ol' Boggy, Lyle was hot on the trail of the truth surrounding yet another fantastic creature: Lizard Man.

As with such famous cryptids as Mothman and Owlman, its name provokes imagery of a comic-book-style super-villain. But, in reality, and as Lyle demonstrates, there's nothing cartoonish about Lizard Man in the slightest. "Nightmarish" is a far better description. So, with that said, what has Lyle served up for us this time?

Lizard Man: The True Story of the Bishopville Monster is an enthralling, chillingly atmospheric, and deeply revealing look at a strange and controversial legend that first surfaced in 1988. This is a story of small-town secrets, of a cast of fascinating and diverse characters, of media hysteria, and of a terrible, and terrifying, animal that just may be more than the myth that many assume it to be.

It's important to note, too, that even though the story is more than a quarter of a century old, its fascination endures—within Bishopville itself, within the mainstream media, and within the field of cryptozoology, as Lyle skillfully demonstrates.

What I particularly enjoyed about *The Beast of Boggy Creek* is that Lyle didn't lazily sit at home, while drawing all of his data from the internet. Instead, he hit the road and sought out the truth, firsthand. And that's precisely what Lyle did while researching the story you are about to read.

With his research partner, Cindy Lee, Lyle headed off—road-trip style—to South Carolina, where he tracked down the key players in the saga (or, at least, those who have not fallen foul of what may be nothing less than a sinister "Lizard Man Curse"), and he explored the wild woods and the spooky swamps of Bishopville, with just one thing on his mind: finding the answers to the beastly puzzle.

It wasn't as easy as it might sound, however. Not only have 25 years passed since the Lizard Man first surfaced, but, as Lyle learned and as he demonstrates, there have been a lot of misconceptions about the reports, the eyewitnesses, and the creature itself.

Rather like Sherlock Holmes and Dr. Watson roaming the foggy wilds of Dartmoor, England, in pursuit of the Hound of the Baskervilles, Lyle and Cindy undertake far more than a bit of dicey detective work as they make their way around town and the surrounding, woody, swampy landscape. And also like Holmes and Watson, our dynamic duo uncover a wealth of untapped data and cases, encounter a variety of people who may be the key to parts of the saga, and realize that Bishopville really does appear to be the domain of something wild, something primitive.

Lizard Man is an important piece of work.

—**Nick Redfern** is the author of many books, including *Monster Diary* and *There's Something in the Woods*.

Introduction

I must admit that years ago when I first came across reports of a "lizard man" that allegedly stalked the swampy waters near Bishopville, South Carolina, I was dubious. The idea of a green-skinned, three-toed reptilian humanoid slinking across roads at night, attacking stranded motorists, and gnawing on car bumpers seemed more like a *Scooby Doo* episode than anything remotely real. Not to say that a real-life Creature from the Black Lagoon isn't an intriguing notion to a monster aficionado such as myself, but when it comes to the likelihood that this might be an actual living, breathing creature, I felt it stretched the boundaries of biology and logic to the limits.

I'm sure I'm not alone in this sort of reaction. The mere name "Lizard Man" can easily conjure images of a cartoonish creature with an extended jaw, scaly skin, and a long tail that drags behind as it leaves its lair in search of some poor soul to scare. It doesn't help that the media perpetuates this image either. News articles and internet pages that recount sightings of the Bishopville Lizard Man often inflate the caricature by including childish drawings of rampaging lizards or walking alligators in their reports. In some cases they've even used photos of toys when no suitable drawing was available. Merchandisers have also followed suit, capitalizing on the legend by selling T-shirts with screenprinted images of goofy reptile men in a variety of primary colors. No doubt this is all in fun, but does it really represent what people claimed to have seen on those dark nights down in Scape Ore Swamp?

It wasn't until I began taking a closer look at the case that I realized perhaps my initial impression of this so-called reptilian humanoid might just be wrong. Following the release of my first book, *The Beast of Boggy Creek*, I was tossing around ideas for a second book. I've always been partial to creatures that inhabit swampy areas—as the Boggy Creek creature does—so I began to take a closer look at some of the others. There are certainly many types sloshing around the realms of cryptozoology, but the one that kept coming up more than any was the Lizard Man of Lee County—a.k.a. the Bishopville Lizard Man. As a cryptid, the Lizard Man shares company with heavy-hitters such as Bigfoot, Yeti, Loch Ness Monster, and Chupacabras, but still manages to carve out its own reputation, which is quite famous among those who follow these sort of real-life "monster" reports. As such, I thought the creature might be worthy of a second look.

So I began to dig deeper into the case. I tracked down countless newsprint articles covering Bishopville's alleged Lizard Man, watched television documentaries on the subject, and asked my colleagues for their opinion. I searched for other cryptozoology cases where some type of reptilian or amphibious humanoid had been described. I even made the 14-hour road trip from Dallas to Bishopville so I could interview the people involved and get a better understanding of the place where the creature was said to dwell. In doing so, I discovered that there was much more to the story than I first realized.

First, the descriptions given by the witnesses didn't jive with the cartoonish media portrayal of a green, scaly alligator man. Their accounts described something strange, but not necessarily outlandish or totally implausible, and in many cases, not even reptilian. As well, the case involved law enforcement officials who took the situation very seriously, providing a wealth of affidavits and documentation that is not always available in these type of investigations. And last but not least, the case also had some fascinating social aspects which were interesting in themselves. The whole thing was not, as I had initially believed, a matter of a few loosely connected incidents. This was a full-blown phenomenon that hit Bishopville like a bomb. There were multiple sightings, major national news coverage, mass merchandising, monster hunters, gossip and rumors, a baffling series of "attacks" on cars, and even a call from the famous news anchor, Dan Rather, as he and the rest of the world clamored to know more about Bishopville's elusive Lizard Man during the summer of 1988.

And things didn't stop there. After 1988, the sightings continued. These later sightings, which are routinely overlooked or altogether absent from most accounts of the Bishopville Lizard Man, are just as baffling as the ones that initially set the frenzy in motion. Those bizarre instances of cars being found "chewed up" have also continued right up until the present day!

The more I learned of these details, the more I realized that the whole story had yet to be told. The Lizard Man is often mentioned in general cryptozoology books or profiled on monster-related websites and television shows, but the majority of these fall short with brevity, errors and/or copious omissions that are perpetually passed on as these sources are referenced elsewhere. And what about Scape Ore Swamp where the creature is said to live? What's it like? Who are these people in Bishopville who claimed to have seen this mysterious creature? These are questions that need to be answered so this case can finally be judged on its own merit and determined to be credible or simply classified as a modern-day urban legend. In either case, the Lizard Man story has always been, and still is, an extremely compelling one that's far overdue to receive its own definitive treatment.

So here we are. What you're now reading is the tale of my journey to find the truth about the Lizard Man. It's not an effort to ultimately prove its existence, but a quest to strip away the scaly layers of media mockery and hearsay, and document once and for all what the citizens and police force of Bishopville have been telling us all along: that there may not be some hidden gem of dinosaurian evolution sloshing around down in Scape Ore Swamp, but by all accounts there is something very strange lurking out there. The witnesses are convinced they've seen it and the local law officials are backing them up. This is their story.

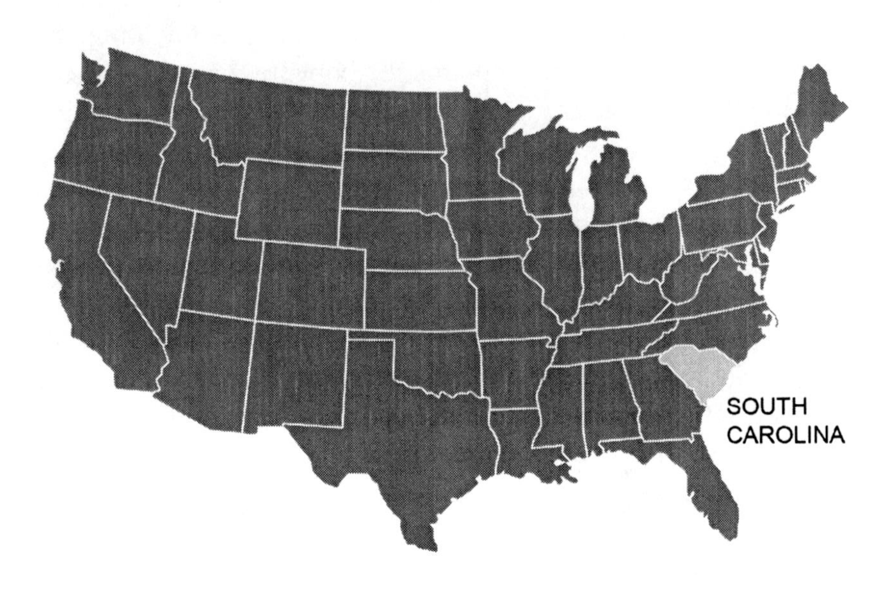

South Carolina Map

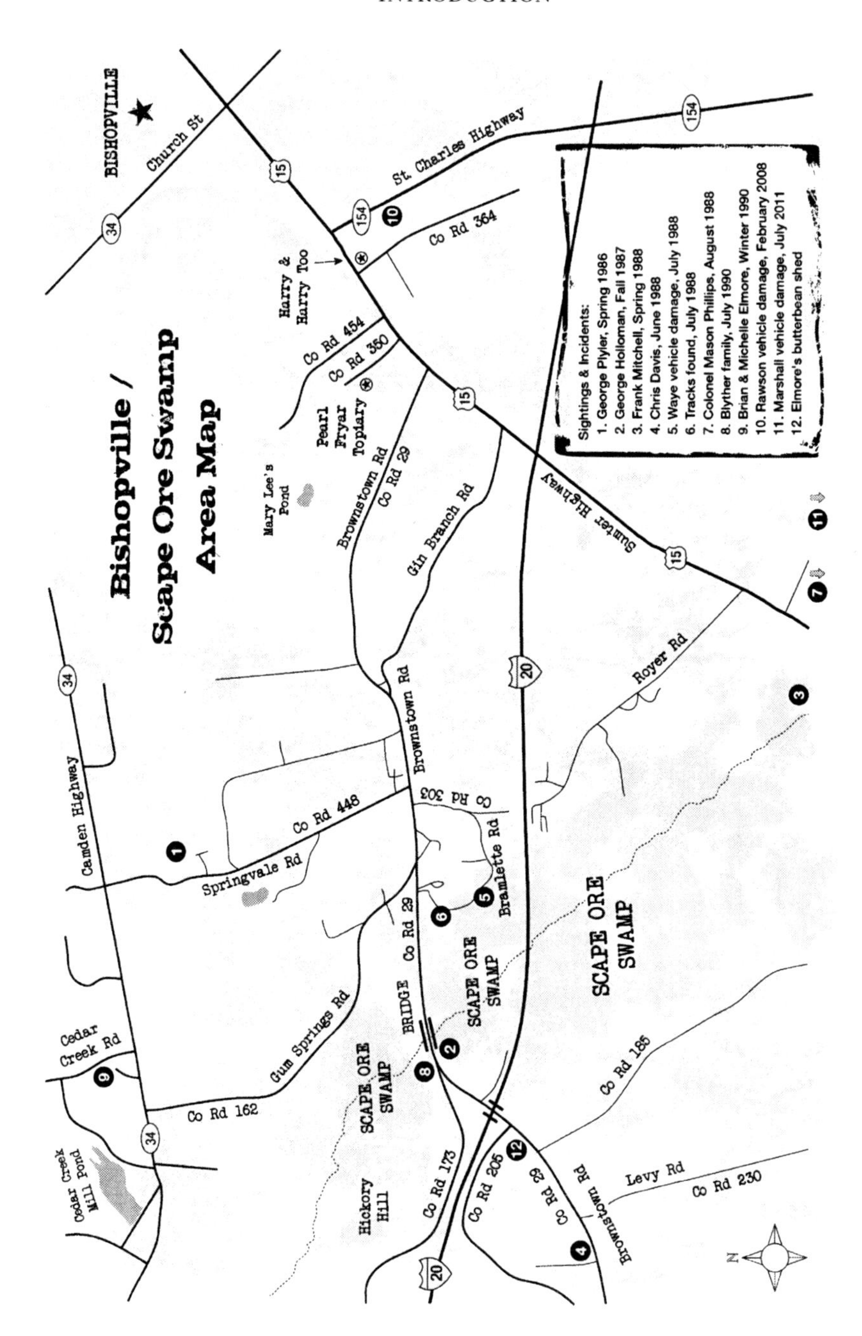

Bishopville Area Map

1. The Lizard Man Cometh

Darkness had already descended on the little town of Bishopville, South Carolina, by the time I turned the car off the main road onto a sleepy side street. I rolled down my window to let in the balmy summer air as I checked the GPS again. Almost there; only one more turn before the final destination.

My research partner, Cindy Lee, and I were headed to the house of a retired sheriff by the name of Liston Truesdale. Truesdale is intimately connected to the Lizard Man story, having been there from the beginning back in 1988, so he was a logical starting point for our investigation. I had contacted Truesdale several weeks before to arrange for the visit, so he was expecting us that late Sunday evening. Sort of.

Since I wasn't sure what time we would arrive, I agreed to call once we got close to the area. I dialed his number several times, but each time I was greeted by the brash pulse of a busy signal. In the age of cell phone immediacy, I'd almost forgotten what a busy signal sounded like. This was old school; a landline. I looked at Cindy with an exaggerated face of concern. I was only half joking, but admittedly apprehensive that we could not get a hold of our host. I had come 14 hours from Texas on the basis of a "Lizard Man." Cindy smiled and reassured me that I wasn't completely crazy for doing such things.

After nearly an hour of busy signals, we figured it might be best to just head on over to the sheriff's house. Perhaps he was just enjoying a long Sunday evening conversation or the phone was simply off the hook. I didn't want to intrude without having made the phone connection first, but I wasn't sure what else to do. Our schedule didn't allow for much leeway, so it was imperative that we make the most of our time.

As directed by the GPS, I made the final turn onto Truesdale's street and slowed down so we could scan the house numbers. The houses, most of which were modest frame structures, seemed to eye us warily as we crept along. The neighborhood was nestled within clusters of huge trees making it shadowy and difficult to see the addresses.

Despite the difficultly, we eventually narrowed the search to a corner lot at the end of the street that corresponded to the address Truesdale had given me. The house appeared completely dark with no visible lights inside or out, making us wonder whether or not anyone was even home. Regardless, I turned the corner and rolled into the driveway, which led to a garage at the back of the house. Through the open car window I could hear the sound of pecan shells cracking under the weight of the tires. I pulled forward a bit more and killed the engine. The night went silent.

After a few moments, Cindy and I got out of the car. Near the driveway, we could now see a side door with a faint light emanating from its glass, but I figured I would try the front door first. Cindy waited while I walked to the front of the house. I started to ring the bell, but somehow it seemed less welcoming, so I decided to forget the formal entry and return to the dimly lit side door. Cindy and I ascended three small steps and knocked.

A minute passed with no response. A dog wailed in the distance and I began to feel like the character in *The Mothman Prophecies* movie who was forced to knock on a stranger's door in the middle of the night after his car broke down. There was a certain surreal quality about the evening… but I suppose that's to be expected when driving around the Deep South in search of Lizard Men.

Finally, a porch light flashed on and we could hear the metallic sound of locks clicking. It was as if the sleepy house had suddenly awoke. A few seconds later the door swung open and an older gentlemen greeted us in a smooth Southern accent. I recognized the distinctive voice; it was no doubt our host, Mr. Truesdale.

Truesdale welcomed us inside and apologized for the busy phone. He'd been talking to a relative for quite some time. His wife had passed away earlier in the year, so the phone had become a respite from the quiet, lonely nights. That and his dogs. Truesdale and his wife had always been dog lovers and their rather large backyard housed kennels for nearly a dozen furry canines.

After introducing Cindy to Mr. Truesdale—and fielding a couple of questions along the lines of "why would such a beautiful young lady want to come searching for *monsters*?"—he invited us to take a seat in his den.[1] The room was spacious and inviting with a rustic fireplace on one wall and a large set of ornate wood shelves built into another. The shelves contained various mementos from his law enforcement career, personal photos, and even a few Lizard Man related items.

Cindy and I settled in on the couch as we chatted with Truesdale about his daily life now that he was a widower. Not surprisingly, much of his time was dedicated to chasing down and signing legal documents to finalize his wife's passing. They had no children, so the daunting task rested solely in his hands. We got the feeling he was overwhelmed but that perhaps he was used to such things. In his time as sheriff, he was often confronted with piles of paperwork and unpleasant phone calls, including those having to do with the Lizard Man case.

It wasn't long before Truesdale steered the conversation towards that very topic. It was late in the evening, and we had merely planned on meeting long enough to strategize plans for an interview and a tour of the area, but the retired sheriff seemed determined to launch right into his accounts of the infamous creature.

I quickly got out my recorder and notepad as he began to talk. The tone of his voice and the reflective pauses in his words made it clear that he still took the case seriously, but we could also sense an underlying tone of frustration. He'd been connected to the saga of the Lizard Man for 24 years now, whether he liked it or not. As a man who had been unwittingly cast into the limelight of a monster tale, he had handled it well. But as a lawman, it was obvious that he wished he could have conclusively solved the mystery, one way or another.

As Cindy and I sat in Truesdale's cozy den, listening intently, I

1 He did ask why I was interested in "monsters" as well, but it wasn't relative to my outward signs of beauty… or lack thereof. I suppose I look the part.

smiled knowing that I had been given a rare opportunity to study this mystery from the inside. Truesdale was the real deal; a legendary figure that was akin to a real-life Carl Kolchak, the Night Stalker, hunting for a creature that defied rational explanation. But this was not a television series or a movie scripted by writers, this was reality. Real people had claimed to see something that no one could account for, and Truesdale was the starting point for uncovering all of their stories. Scape Ore Swamp, where it all began, was somewhere out there in the darkness only a few miles away, still clutching its secrets within the vast, murky waters. No matter what I would ultimately conclude about the Lizard Man mystery, I could already tell it was going to be fascinating journey.

Waye Out There

On the morning of July 14, 1988, the Lee County Sheriff's Office received a bizarre call. A family by the name of Waye reported that their 1985 Ford LTD had been "mauled" during the night while they slept. The vehicle, which had been parked under an open metal carport, had suffered extensive damage to the molding, sidewalls, and hood. It would have seemed like the work of a vandal, but hair and footprints found on the car led them to believe some kind of animal might have been responsible. But what kind of animal would sink its claws and teeth into the metal skin of a car? Two deputies were dispatched to find out.

The deputies made their way to the residence, which was located in a small rural community known as Browntown on the outskirts of Bishopville. When they arrived, homeowners Tom and Mary Waye and their daughter Shirley showed them the vehicle in question. Just as they had reported, the chrome molding had been torn away from the fenders, the sidewalls were scratched and dented, the hood ornament was broken, the antenna was bent, and even some wires from the motor had been ripped out. Upon closer inspection, it appeared that parts of the molding had actually been *chewed*, as if an animal had used its teeth to inflict the damage.

To further support the animal theory, the Wayes pointed out clumps of reddish colored hair and muddy footprints that had been left all over the car. Their home sat on the edge of a vast and rugged region called Scape Ore Swamp, so it would not be out of the ordinary for a wild animal to pay them a visit. Except, of course, the animals didn't usually resort to acts of vehicular vandalism.

Given the unusual circumstances, the deputies decided it might be best to call Sheriff Liston Truesdale to the scene. Truesdale was a seasoned lawman who had been in office since 1974. Over the years, he had seen just about everything the local area could throw at him, so perhaps he could make some sense of it. But neither the deputies nor Truesdale could have imagined the impact this call would make on their lives.

"This is what started the whole thing," Truesdale told Cindy and I, as he recalled the sequence of events. "The deputies that were dispatched called back and said, 'This is kind of unusual. You better come out here and take a look.'"

Truesdale promptly made his way to the scene. Once there, he assessed the situation by examining the evidence and interviewing the Wayes himself. Upon hearing their story, he felt they were being truthful in the matter and agreed that perhaps an animal was the culprit, however unlikely it may have been. To be certain, however, he would need a more qualified opinion, so a biologist from South Carolina's Wildlife and Marine Resources Department was summoned to the home. The biologist examined the hair fibers and surmised they could belong to a red fox, although he could not explain how a small canid would be able to cause such damage. Under the circumstances, he decided to collect samples of the hair so they could be sent to the University of Georgia for further examination.

Truesdale and the biologist also examined the footprints left by the animal— or animals—and concluded they were mostly likely those of a fox. Several larger tracks were also found approximately 25 yards from the car, headed towards the swamp. These appeared to be from a large biped or quadruped, possibly a bear. Black bears had occasionally been seen in Scape Ore Swamp, so again, this would not be completely out of the question.

It was fairly clear that the tracks on the car were those of a fox, but not everyone agreed that the larger tracks were those of a bear. A local man, James Knight, who also examined the tracks, found them to be somewhat unusual. In a 1988 interview with *The Item* newspaper out of Sumter, South Carolina, Knight was quoted as saying: "The tracks were aligned in a straight procession typical of human steps."

Needless to say, everyone was completely baffled by the incident. If the Wayes were telling the truth, then it only left a few possible explanations: 1) The car had been damaged during the night by human vandals, who either left fox hairs and paw prints as a red herring, or had been followed by a curious animal who jumped on the car and perhaps chewed some of the detached molding after the vandals had fled. 2) A fox, bear, and/or some other animal had damaged the car—perhaps while fighting each other—which would include crawling up into the engine area *or* opening the hood to tear the wires! Neither explanation made much sense, but it was all they had.

By now a growing number of onlookers had gathered at the scene. Many of them were neighbors who lived in the area, but others were individuals who had heard the strange dispatch call on their private police scanners and had made their way to Browntown out of curiosity. Even a few reporters had gotten wind of the incident and were showing up with notepads, recorders, and cameras in hand.

Truesdale was trying disperse the crowd and finish his investigation when some of the locals informed him there might yet be another, more bizarre possibility. "While we were there looking over this situation, we learned that people in the Browntown community had been seeing a strange creature about seven feet tall with red eyes," Truesdale told us. "Some of them described it as green, but some of them as brown. They thought it might be responsible for what happened [to the car]."

The sheriff and his deputies were taken aback. Were people suggesting that some kind of *monster* was on the loose? It seemed

The Waye's damaged vehicle (photos by Liston Truesdale)

completely far fetched, but the seriousness with which the locals told the story compelled the lawmen to consider the possibility.

"Naturally, when we heard this, we started trying to find out more," Truesdale recalled. "We began to ask for the names of anyone who had seen such a creature. If they were telling the truth, then surely somebody would know something." And sure enough, they did. The officers were given the names of several people, one of whom had supposedly encountered the creature one evening near the Scape Ore Swamp bridge. The bridge was located near the Waye's home, so it seemed like a promising lead, even if it turned out to be a hoax or some other mistake.

Truesdale jotted down the names and headed back to his patrol car, satisfied that he had done all he could for the moment. It would take time for the hair analysis results to come back, plus he simply needed some time to digest the whole situation. It had been one the most bizarre police calls he had ever received.

But there was one more surprising revelation in store.

"As we were leaving the house, I saw a guy that I knew by the name of J.J.," Truesdale recalled. "I said 'J.J., have you heard anything about a tall creature, possibly green with big red eyes?' He said, 'What do you mean, that *Lizard Man*?' And that's how we first heard of it."

The Lizard Man. It was a simple name, but one that conjured an array of wild mental images. It was also a name that would soon come to the attention of the public by way of news reporter Jan Tuten. Tuten, who at the time wrote for South Carolina's largest newspaper, *The State*, happened to be riding with Truesdale when J.J. made his revelation. She had been at the Sheriff's Office when the initial call came in from the deputies and had asked if she could ride along. Now, like Truesdale, she found herself trying to make sense of the peculiar situation.

Truesdale thanked J.J. and returned to the station. He was skeptical that a monster was lurking in Scape Ore Swamp, but decided it would be best to take the situation seriously until he could get to the bottom of it. Over the next few days, he and his deputies set about trying to locate the witnesses whose names they had been given. To accomplish this, they spread word that if anyone had any information about the creature itself or the vehicle damage, that they should contact the Sheriff's Office. Nobody was in trouble, they just wanted to talk.

"The whole thing was way out there, but it was our duty to follow up. So that's what we did," Truesdale told us emphatically. Unfortunately, the police had no luck either in finding these people or getting them to come forward. It was clear that people in Browntown were aware of a strange creature, but when it came to sharing eyewitness information with the law, things seemed to have reached a dead end. That would all change on the afternoon of July 16, however, when a man brought his son to the police station. The boy's name was Christopher Davis and he had quite a story to tell.

THE ATTACK

Sheriff Truesdale was going about his usual day's work at the Lee County office, when Tommy Davis walked in with his 17-year-old son, Chris. Mr. Davis had read an article about the Waye's vehicle damage in the newspaper and felt that he might be able to help. Apparently, Chris had encountered some kind of strange creature two weeks earlier that may well have been the reputed Lizard Man. The young man had been frightened so badly during the incident that his father just couldn't let it go any longer without telling the police. Especially if there was a dangerous creature on the loose, as the car damage seemed to imply.

"I was there that afternoon by myself when Chris Davis' daddy brought him in," Truesdale told us, as he laid out the circumstances surrounding the most famous Lizard Man encounter. "Chris started telling me about what happened and I couldn't believe it. I mean, it was so far out."

The incident took place in the early morning hours of June 29, Chris said. He had been working the night shift at a McDonald's restaurant and was on his way home at around 2:30 a.m. The route took him along Browntown Road, which crossed Scape Ore Swamp in a thickly wooded area. Approximately a mile past the Scape Ore bridge, he heard a loud pop as his '76 Celica began to jitter with the tale-tale signs of a flat tire. Cursing his luck, the teen pulled to the side of the two-lane road and stopped the engine. He got out and looked at the tire. He didn't have a flashlight, but there was enough moonlight to confirm that it was indeed flat.

Chris was parked at the intersection of a small dirt road that led into the surrounding fields of cotton, high grass, and trees. There were no houses for at least a half mile. He didn't exactly feel like changing a tire on the lonely stretch of road, but his own house was still seven miles away so walking was out of the question. Faced with no other choice, Davis opened the trunk and pulled out a tire iron, jack, and a spare. He wanted to get it over with. The bugs were already starting to eat him, and the fish sandwiches he had brought from work were getting cold in the passenger's seat.

Davis swapped the tire as quickly as he could and gathered up his tools. As he was placing them back in the trunk, something

The location where Chris Davis said he pulled over to change the flat (photo by the author)

caught his eye about 30 yards away. It was moving towards him, swinging its arms. In an interview printed in *The Item* newspaper on July 20, 1988, Chris recalled that: "I had just put the tire in the trunk when I see this thing coming from the trees." He described it as a large, humanoid-like creature with "glowing red eyes." It was "green, wetlike, about seven feet tall and had three fingers," he told the press. Or in simpler terms, "a red-eyed devil."

The thing was getting closer by the second. The young man panicked. "I ran to the driver's side and got in," Chris recalled. "When I was sitting in the car, I saw him from the neck down. He was outside the driver's side window. After about two yards, he jumped on the roof. I saw hands, rough-looking, black fingernailed hands, sticking down from the windshield!"

The creature then let out a deep grunt and that was enough to send the teen into action. He turned the key, fired up the Celica, and stomped on the gas pedal. As the car lurched forward, the creature fell off, but managed to get up and give chase. Chris estimated that he was going at least 35 mph when the creature finally caught up and attempted to leap on the car again. "I looked in the rear-view mirror and saw something. And I heard a crash on the roof."

Chris was truly in fear for his life by this time. He began to swerve back and forth, trying to dislodge the attacker. He never saw it fall, but eventually the swerving must have worked because the creature was no longer clawing and banging on the roof. At that point, he held the accelerator down and sped for home.

When he pulled into the driveway, Chris blew the horn frantically before jumping out of the car and running into the house. His father told reporters that "[Chris] was so upset he left the car running. He was huffing and puffing. In a few minutes, he started crying. We asked him what had happened, and he said what he saw was 7 feet tall, had red eyes and three fingers on each hand."

Shocked, Mr. Davis went out outside to inspect the car. He noticed that the driver-side mirror was bent and twisted. He also checked the roof and found scratches in the paint. Was this evidence of the creature's attack? It was all very strange, but he felt sure that something had truly happened to his son. The car damage, along

Chris Davis encounters a terrifying creature

with his Chris' state of shock, seemed to confirm it.

Tommy Davis was well-respected in the community, having worked for years as head of the electrical department at a large plant in Sumter, so Truesdale felt that he was trustworthy. Chris also came across as genuine during the police interview. Even though his story was outrageous, Truesdale's gut instinct told him the young man had truly seen something unexplainable that night. He just wasn't sure *what*.

Chris and his father attested to the fact that Chris had not been drinking that night. After all, he had just come from work. Chris also told the sheriff he had heard stories about some kind of "weird creature" going back at least two years, but he never thought too much about it. He didn't discount the possibility that it could have been a bear with wet, green mud covering its fur, but its actions just didn't seem like something a bear would be capable of doing. He never claimed it was a "lizard man," only that it was strange, frightening, and like nothing he had ever seen before.

Sheriff Truesdale interviews Chris Davis
(photo courtesy of Liston Truesdale)

After having Davis recount his story a second time on tape—which Truesdale noted was exactly the same as the first time—the sheriff asked the teenager to sketch what he had seen. Davis grabbed a pencil and proceeded to make a crude sketch of the alleged creature on a piece of notebook paper. The drawing lacked any sort of artistic flair, but clearly showed an upright, humanoid figure with three prominent fingers on each hand.

Truesdale pondered the situation for a moment, then asked if they were willing to undergo a lie detector test. Chris and Tommy agreed without hesitation. The sheriff then dismissed the pair, telling them he would get in touch soon to set up the polygraph session.

Alone again in the office, Truesdale was left to consider what had just taken place. He had asked witnesses to come forward with any knowledge of the so-called Lizard Man, to which Davis explicitly complied, but now the sheriff was not so sure he liked the outcome. The weird car damage had been one thing, but now he had a person claiming to have come face-to-face with some kind of hostile humanoid creature. It was as if a nightmare had walked out of Scape Ore Swamp and grabbed him by the badge. He was sworn to protect and serve the citizens of Lee County, but what exactly was he protecting them from? A joke or a living, breathing Creature from the Black Lagoon? He could not be sure, but either way, he felt it was best to keep an open mind for the time being and allow solid police work to sift out the truth.

"We had to keep looking into the situation at that point," Truesdale told us during our interview. "If it turned out to be nothing, then fine. But if it turned out to be something dangerous, then we would've been in real trouble if we hadn't done something about it."

Large Scale Menace

Once the details of Chris Davis' alleged attack went public, it was as if a scaly bombshell had hit the community. Chris was not only the talk of the town, he was immediately deluged with requests for interviews by local reporters. Newspapers, which had initially run stories about the Waye's vehicle damage, seized upon the new, sensational details of the Chris Davis incident and worked quickly to get them in print.

Radio stations also got in on the action. WAGS-AM, a local Bishopville station, was the first to broadcast news of the Lizard Man. According to the July 21, 2004, edition of *The Item*, station owner Emory Bedenbaugh got the scoop from police. "I had heard some rumors about this creature that was observed in Browntown, and I went over to the sheriff's office to ask about it," Bedenbaugh told reporters. "I remember Deputy Chester Lightly saying to me, 'Oh they call it the Lizard Man.'" Bedenbaugh returned to the radio station and began to talk about the Lizard Man on the air. He also called the South Carolina News Network, the Associated Press, and United Press International.

The Lizard Man stories enthralled the public, who reacted with a mix of skepticism, fascination, and curiosity. Carloads of curious sightseers began to jam Browntown Road as they came for a glimpse of Scape Ore Swamp and, if they were lucky, the creature itself. "Traffic on the road was steady from morning to midnight, something local woodsmen say is a direct result of the sightings," *The Item* newspaper reported on July 20.

To make things worse, the WCOS-FM radio station out of Columbia offered a million dollar reward to anyone who could "bring in the lizard thing alive." This not only attracted more outside attention, it also brought in scores of would-be hunters, who were often armed and dangerous. According to *The Item* article: "Four local friends, each about 20 years old, brought rifles and trucks to the bridge in an effort to do a little sight-seeing and maybe collect the reward. When asked if he would try to spare the creature and thus meet the main requirement for collecting the money, Tony White replied, 'Damn straight, I wouldn't have to work a day in my life.'"

Reporters, television crews, and photographers also began frequenting the area in growing numbers, as they fished for information from the locals and shot rolls of film to accompany their articles and features. The July 20 edition of the *Press Courier* reported that: "The swamp was swamped with TV crews and other curious people hoping to catch a glimpse [of the creature]."

Back at the police station, Truesdale and his deputies were neither laughing nor grabbing for there guns. They were too busy trying to deal with the influx of calls coming in from the press. Much of their day was being spent answering silly questions and repeating the same stories over and over, which took time away from the actual case, and more importantly, time away from more critical police work. The Lizard Man phenomenon was barely a week old and it was already getting out of hand.

But this was only the beginning.

Around 2:00 a.m. on July 24, two teenagers rushed into the Lee County Sheriff's Office and asked to speak to an officer. They were breathing hard and obviously upset. Deputy Wayne Atkinson responded. As he approached the teens, he asked what the problem was. Without hesitation they blurted out: "I think we saw the Lizard Man!"

The two teens, Rodney Nolf and Shane Stokes, said they were out driving with their girlfriends along Highway 15 when something darted across the road in front of them. It appeared to be "a large, muscular animal," which moved on two legs. The creature was only visible for a few seconds before it jumped a tall fence and disappeared into the woods. Deputy Atkinson could not be sure it had anything to do with the so-called Lizard Man, but he felt they were telling the truth based on their sincerity and demeanor. Atkinson told reporters at *The Item*: "I've been in law enforcement long enough to know when somebody's scared."

A few hours later, the Sheriff's Office received another strange report. This time it was a call from a Browntown resident who said she heard some sort of unusual "howls" coming from the nearby woods. The office wouldn't normally have paid much attention to such a report, but given the Lizard Man situation and the earlier sighting, Atkinson thought it might be good idea to drive over to the area and have a look around.

One of the discarded drums found on Bramlette Road (photo courtesy of Liston Truesdale)

State Trooper Mike Hodge was also on duty, so Atkinson convinced him to ride along. The two men drove over to Browntown where they began slowly patrolling the backroads using a spotlight. While driving along Bramlette Road, they came upon several 40-gallon drums that had been crushed and scattered on the road along with some other garbage. They decided to get out and investigate.

The officers first examined the drums but could find no apparent clues as to what had crushed them or why they had been left in the road. While looking around with flashlights, they also noticed several tree limbs, hanging nearly nine feet overhead, which had been snapped. All the while, the officers got the feeling that someone, or some*thing*, was watching them. Heavy woods lined the road, so it wouldn't be too difficult for something to conceal itself just out of sight.

Figuring the garbage must have fallen off the back of a truck or had been dumped on purpose, the officers returned to their car and left the scene. They continued down the road a short distance but saw nothing else of interest, so they made a U-turn and doubled back. As they passed by the strewn garbage a second time, they saw what looked to be footprints going across the road… footprints that were not there a few minutes before!

They immediately stopped the car and got out. There in the dirt was a set of huge animal tracks unlike any they had ever seen. The impressions clearly showed a large oval palm pad, a prominent heel, and three claw-like toes which sunk to a depth of nearly 1½ inches. From toe to heel the prints measured a whopping 14 inches in length, and at the widest portion, 7 inches across. The stride was just as astounding, measuring over three feet per step. The lawmen were not sure what to make of it. Was this the footprint of the elusive Lizard Man or the work of a clever prankster? How could a prankster have known they would drive by the area again?

"I was pretty spooked," Trooper Hodge was quoted as saying in the July 28, 1988, edition of *The Rock Hill Herald*. "They were just some weird tracks. They were too consistent to be fake. They were deep down in that hard dirt."

And indeed, one of the curious aspects of the tracks was how difficult it would have been for a hoaxer to make the foot impressions in such dry substrate, especially with such a short window of opportunity. "I stomped in the road, and I couldn't make a track," Deputy Atkinson added. He was completely baffled.

The officers did not like the idea of having to follow the tracks in the dark, but they knew it would be best to investigate while the trail was still hot. So with one hand on their gun and the other holding a flashlight, they followed the trackway for approximately 300 yards before the footprints finally veered off into the brush. At that point, they decided it would be best to wait for daylight. It was simply too spooky to think about trudging off into the dark woods. Whatever had made the tracks might be waiting for them out there.

Muddy Waters

Sheriff Truesdale was awakened by the sound of his telephone. He answered it with a groggy "hello." It was the office dispatch. Deputy Atkinson was requesting that he come down to Bramlette Road as quick as possible.

"The sun wasn't even up that Sunday morning, when I got the call," Truesdale told us. "They wanted me to come down and look at some unusual tracks they had found."

Truesdale threw on some jeans and a T-shirt and made his way to Browntown just as the sun started to break over the horizon. It looked as though it was going to be a beautiful day, but somehow he had a feeling it might not be as bright for him. He had hoped to sleep in a little before he had to meet with a group of Fox TV reporters from New York later that afternoon. Giving up part of his afternoon to appease the media was one thing, but now he had been forced to give up his Sunday morning as well... all because of the Lizard Man.

When Truesdale pulled onto Bramlette Road, he found Atkinson and Hodge waiting for him. They showed him the strewn garbage, broken tree limbs, and strange tracks. The Sheriff agreed that the situation was highly suspicious and wanted to get to the bottom of it as soon as possible. He had the deputies block off the road while he put in a call to the State Law Enforcement Division (SLED) to request some bloodhounds. He also sent one of his men to retrieve some plaster material so they could make casts of the footprints.

While he waited, Truesdale examined the area more closely. He noted the trees had their branches snapped off at a height of eight or nine feet. It appeared that something large and heavy had come crashing out of the woods, walked a good distance down the road, and then headed back towards the swamp. The three-clawed footprints almost seemed too convenient and too monstrous, but he could find no evidence of any shoe prints or machinery marks which would have instantly exposed a hoax. The circumstances

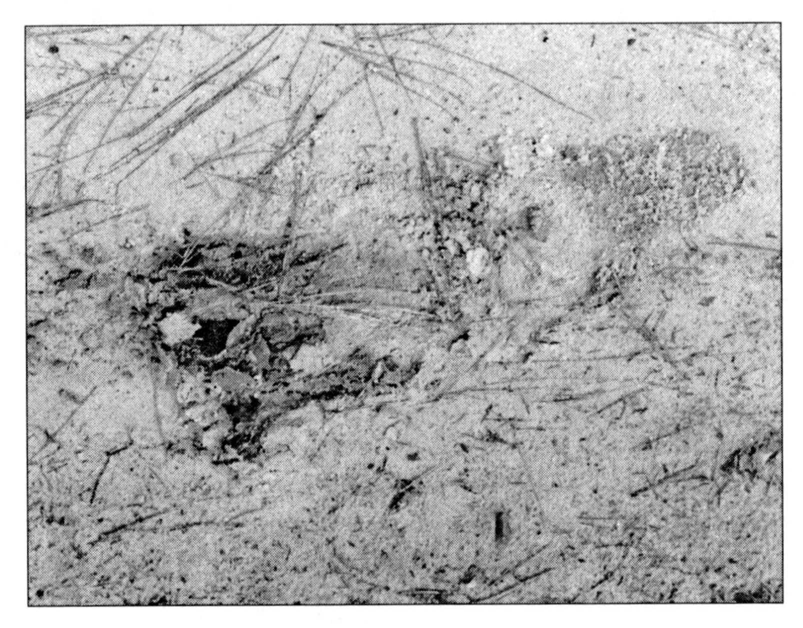

One of the track impressions found on Bramlette Road (photo courtesy of Liston Truesdale)

were as strange as the Waye's vehicle damage.

By the time the deputy returned with the casting materials, the cheerful sky was beginning to give way to grumbling clouds and gusty winds. A summer storm was brewing over Bishopville. The officers eyed the weather as they mixed water and plaster together in a bucket and poured the liquid into five of the best tracks. They hoped the rain would hold off long enough for the material to dry.

Truesdale was speaking with his team when he noticed several cars pull up to the roadblock. "CNN showed up down there as we were investigating," Truesdale chuckled. "They wanted to do a story on the Lizard Man. I don't know how they found us."

Truesdale already had the Fox TV interview scheduled for later in the afternoon and now CNN news reporters had shown up. Since he had been called out of bed on a Sunday morning, he was not even dressed in his uniform. He looked overhead as he walked down the road to meet with them. The grumbling clouds echoed his sentiments for the media circus.

Truesdale approached the reporters with an affable smile, however, and convinced them to wait while he and his deputies finished their investigation. He was willing to do an interview but police matters came first. The reporters were dying to get a better look so they agreed to shoot some B-roll footage while they waited for the sheriff.

In the meantime, three wildlife officers from the State Law Enforcement Division arrived with a team of highly trained bloodhounds. Upon initial examination of the tracks, the officers didn't feel they could be from a real animal but nonetheless agreed to have their dogs search the area for more evidence. The hounds were ushered out of the transport vehicle and given an opportunity to sniff the track impressions. The dogs had a hard time locking onto a scent, so the officers walked them along the road hoping that perhaps they would have better luck as the tracks progressed.

Eventually, the officers and dogs reached the point where the footprints left the road and headed into the trees. The impressions were not as clear in the brush as they were on the dirt road, but even so, they managed to trail them nearly 150 yards into the swamp until the threatening clouds finally unleashed a downpour which quickly, and unequivocally, washed away any remaining chance to find out what may lie at the end of the trail. The wildlife officers were never convinced they would find a monster, but even if they had found something more mundane, it might have helped solve the mystery. Now they would never know.

Fortunately, the deputies had been able to extract the five track castings before the rain hit, so at least they had some concrete evidence. The lawmen entertained the possibility of sending one of the casts to an independent expert for a second opinion, but the wildlife officers convinced them that "there was no way anyone would identify them as a known species of animal," especially since their dogs couldn't even pick up a definitive scent. It was admittedly bizarre but not worthy of further consideration.

The bloodhounds may not have been able to pick up a scent, but Truesdale was beginning to smell a rat. The tracks were highly suspicious, and it *was* possible that someone could have dressed up in a crazy costume to scare Chris Davis. Truesdale never doubted that the teen had been frightened by *something* that night, but he couldn't be sure if it was man or "Lizard Man." None of this explained the sighting on Highway 15 in which a large, muscular creature ran in front of four people, but of course there was no way to rule out the possibility that a bear or some other known animal was to blame for that incident.

If the entire Lizard Man thing was a hoax, it made the situation even more distressing from a law standpoint. This would mean that a person—or persons—was purposefully trying to frighten the community by tearing up cars, chasing stranded motorists, and faking monster tracks. All of this led to sightseers crowding the roads in Browntown, and gun-toting monster hunters potentially endangering their own lives or the lives of others. According to an article published in the July 20, 1988, edition of *The Lee County Observer*: "One sheriff's deputy called to the Browntown section of Lee County near the swamp reported that at least 20 people were out with flashlights—and guns—stalking our reptile." The deputy also noted that "one 'Lizard Man' hunter did actually aim his weapon at a passing bicyclist."

In addition to the hunters and the cars parked alongside the road posing a hazard to passing motorists, there were hidden dangers much more palpable than a Lizard Man waiting in the swamp; these included snakes, sink holes, and fallen branches. Sooner or later, somebody was going to get hurt.

The cops had a real situation on their hands and the sooner they could get to the bottom of it, the sooner it would all just go away. To that end, Truesdale felt it was still important to locate the other Browntown residents who claimed to have seen the so-called creature. If their stories were anything like the one Chris Davis told, he definitely wanted to hear them.

He would not have to wait much longer.

2. The Plot Quickens

"Would you all like something to drink?" Truesdale asked, suddenly breaking away from the Lizard Man conversation. I glanced at the wall clock and realized we'd been talking for nearly two hours. Cindy and I grabbed our half-full water bottles and politely declined.

Truesdale nodded and pointed to an object on the shelf behind him. Most of the items were nicely framed photos, stylish figurines of horses and dogs, or fancy-looking law enforcement mementos, but this was a colorful package containing an action figure. "They sent me that," Truesdale said proudly. "It's the Lizard Man."

I walked to the shelf and picked it up. It was a toy from *The Secret Saturdays* kid's show on Cartoon Network. Sure enough, it said "Bishopville Lizardman" right on the front. Behind the clear plastic stood a 3½ inch figure of a human-like reptile with green skin, webbed hands and feet, and fangs. I turned the package over and looked at the graphics advertising the other "cryptids" in the series.[2] There were six others, but these all had generic names such as "Komodo" and "Piasa Bird." The Bishopville Lizardman was the only character whose name carried a specific location.

I was amazed at the level of notoriety the small-town monster had achieved. In spite of controversy and cynical doubts that had been cast upon the case over the years, the Lizard Man had transcended it all to become legend; a real contender in the realm of Bigfoot and the Loch Ness Monster. The Lizard Man may not be the first to come to mind when talking about cryptozoology, but the story is certainly well known to those with an appreciation of such things, and apparently to viewers of Saturday morning cartoons. This wasn't "the lizardman," this was the "Bishopville Lizardman."

2 A cryptid is an animal whose existence is disputed or unsubstantiated.

It was a distinction that forever linked it to Truesdale's jurisdiction.

I placed the package back on the shelf and took a moment to look more closely at Truesdale's law enforcement certificates. He had become the stuff of legends. Everyone asks him about the Lizard Man—a question he is obviously willing to entertain—but not many ask about the man behind the badge. I returned to the couch and turned the recorder back on. I wanted to know a little more about the man who had so graciously opened up his home to entertain my cryptozoological pursuits.

"I'm also a certified pilot," Truesdale told us, as we talked about his hobbies and interests. He handed over a photo of a red and white propeller airplane. "That's mine," he said proudly.

Truesdale was indeed an interesting man, having a passion for both flying and photography. He dug out a stack of photos and handed it to us. These were aerial views of the Bishopville area. Cindy and I thumbed through them, marveling at the vast green countryside that spread out around the small town. It provided an

Aerial view of Bishopville
(photo by Liston Truesdale)

excellent perspective, one that Truesdale could enjoy anytime he needed an escape from the trappings of pedestrian life and routine patrol.

I asked about his background in police work. He told us that prior to entering law enforcement, he attended the esteemed Southern Police Institute at the University of Louisville. From there he spent 18 years as a Lee County deputy before he was appointed sheriff in 1974. He continued to be re-elected as Sheriff until retiring in 1993. Along the way, he received additional training at the University of Southern California and the FBI Academy at Quantico, Virginia, and was considered an expert in fingerprint analysis and crime scene investigation. All in all, he had dedicated 37 years of his life to fighting crime.

This was an amazing record of service. One that went far beyond any superfluous monster tracking that had come with the territory. Sure, Truesdale had been part of a small police force, but he was a dedicated and truly qualified officer who, for better or worse, found himself at the center of a unique case. It was a rare situation that constantly tested all of his skills (and patience) as it swirled him a vortex of media sensation and bizarre incidents. He was doing his best to make sense of it all, but the strange stories just kept coming.

More Come Forward

Several weeks had passed since the strange footprints were found. During this time, some kids came forward claiming to have also seen strange animal tracks in the snow earlier that winter. They had no supporting evidence, however, so it did nothing to help the case.

The next report of substance came from a man whose name had first come up while officers were investigating the Waye's car damage. It had taken a while, but he finally came forward admitting that he had seen something that might be of interest to the police. The witness this time was a 31-year-old Browntown resident and construction worker by the name of George Holloman. In an official statement taken by the sheriff, Holloman told how he had been frightened by some kind of creature near the Scape Ore bridge approximately 10 months earlier.

On that evening, Holloman was riding his bicycle on the quiet backroads of Browntown. It was early October, but the fall chill had not yet set in, making it warm enough to work up a thirst. Around midnight, he pedaled over to the south side of the Scape Ore bridge where a natural artesian well flows from a small pipe. He parked his bike and took a long drink of the cool water.

It was a pleasant moonlit evening, so Holloman was in no hurry. He got a second drink and then lit up a cigarette. As he stood there smoking, he gazed across the road toward the other side of the swamp. Something caught his eye. According to the official police report, "he saw what he thought was a dead tree that lightning had struck." However, he soon realized it wasn't a tree when the dark shape "stood up like a man." It was "7 to 8 feet tall," black in color and possibly covered in hair. Holloman dropped his cigarette and stood motionless as he watched the thing eyeing him. He told police: "The hair stood straight up on my head and all the muscles got tight in my arms."

The creature, or whatever it was, continued to look at Holloman until the sound of an approaching car broke the silence. As the car passed by, its headlights reflected in the creature's eyes, causing them to glow. Holloman could see that they were big and red. The creature, presumably startled by the car, then moved further into the pitch black swamp until it disappeared from sight.

Holloman immediately jumped on this bike and pedaled furiously until he got home, never looking back. "I didn't hesitate. It went in the swamp and I went thata way," he stated.

Back home, Holloman contemplated what he had seen. He had not been drinking that night and was certainly not hallucinating. Whatever he had encountered there was flesh and blood; something unlike he had ever seen before.

Two weeks went by before he told anyone of the incident. He was reluctant at first, figuring they would think he was drunk or "going fool," but after a while he finally told his family. His brother

Scape Ore bridge where Holloman allegedly saw the Lizard Man (photo by the author)

playfully accused him of being drunk and his mother suggested that he had seen "a haint"[3] (ghost), as if that was anymore comforting. Holloman didn't care if it was a ghost, Bigfoot, or the so-called Lizard Man. He was not about to go back and find out.

When he came forward to tell police of the incident, Holloman agreed to give a full written statement, which Truesdale took on August 13, 1988. Truesdale's impression of Holloman was that he was an honest person who was telling the truth as he knew it. Since Holloman had been so reluctant to come forward and did not seem particularly interested in media attention, Truesdale thought this also spoke well of his credibility.

In the years following his sighting, Holloman did grant a few television interviews. One of the more notable interviews appeared in the Australian documentary show, *Animal X: Natural Mystery Unit*, as part of its "Reptilian Creatures" episode, which aired on the Discovery Channel. In the interview, Holloman recounts his

3 *Haint* is an old Southern term which can mean "ghost" or "haunt."

A strange creature is seen at the Scape Ore bridge

experience, which has not wavered since he first reported it to police in 1988.

* * *

Not long after Holloman came forward, Truesdale located another local resident by the name of George Plyler, who felt he had seen something out of the ordinary. Like Holloman, Plyler had never told anyone outside of a few friends, but was willing to tell Truesdale if he really wanted to know.

In an official police statement, Plyler estimated the incident occurred in the spring of 1986, more than two years before Chris Davis met his monster. He and a few other men were working around daybreak near his home on Springvale Road where the high ground met the boggy bottoms of Scape Ore Swamp. In an interview with *The State* newspaper, he described the eerie event: "I had three workmen putting a hog pen in the swamp in quicksand that came up to your knees and waist. We had gotten out to the middle when I felt the hairs on the back of my neck raise up like someone was watching me. I turned around, and I thought it was a deer glancing out around the tree."

But it wasn't a deer; it was something else… a creature he could not immediately identify. For a few moments the animal eyed him from around the thick tree trunk. Plyler could see that "its face was shaped like a human's, its eyes were red around the pupils, and its arms hung like an ape's." Seconds later the animal took off running on two legs, like a man. "It broke and ran, parting the brush both ways," Plyler explained. At that point he got a better look. During an interview for the *Animal X: Natural Mystery Unit* television show (the same one that interviewed Holloman), he described the creature in detail: "The features were more like a human than any animal I've ever seen. The arms were a little bit longer than normal human-being arms, but the legs were long and skinny, along with a little body." When asked if he thought it resembled a lizard, Plyler explained: "It had a body maybe similar to a lizard except for the head, and it didn't have a tail or anything like that."

After the creature disappeared into the murky swamp, Plyler

glanced over at the other men. They were still working, apparently oblivious to the whole incident. Plyler thought for a moment and decided to keep it that way. He told reporters from *The State*: "I couldn't say anything because those workmen never would have finished the job. They would have told others, and no one would have helped me finish it."

And he was probably right. At the time Plyler was working in the swamp, nobody had been exposed to news of a fantastical "Lizard Man." What he saw was simply a very weird creature that defied rational explanation. The creepy incident was further underscored when he later examined the area where the creature had been standing. There he found odd "three-pronged" tracks in the soft mud. Although he didn't have a ruler, he estimated their size to be 12 to 18 inches on the average.

The experience did not sit well with Plyler. He was sure of what he saw, but was not sure what to make of it. "I didn't tell anybody for quite a while because I didn't want people to think I'd lost my mind," he told the host of *Animal* X. "I know that I'm truthful and honest, one hundred percent. I'm a Christian and I go to church. I've got a college education, two years of business college, and I know what I saw. It's in my mind; it's imprinted and it'll be there when I die."

George Plyler passed away on October 27, 2009. His story never changed.

Assassination Attempt

With the growing number of witness reports and the steady stream of would-be hunters combing the wilds of Scape Ore Swamp, it seemed only a matter of time before the elusive creature might find itself on the wrong end of a gun. And that's just what happened, according to a local man military man by the name of Kenneth Orr. On August 6, 1988 *The State* newspaper splashed the tantalizing headline across the front page: "Florence man says he wounded 'Lizard Man.'"

Orr, who lived in nearby Florence, was an airman assigned to Shaw Air Force Base located 30 miles south of Bishopville. According to a statement given to Sheriff Truesdale, he was driving to the base at 6:00 a.m. on August 5 when he exited Interstate 20 at Highway 15. As he neared Gin Branch Road, he saw a strange creature running down an embankment. He described it as being "green, but not scaly and slimy" with a "lizard-like tail" and approximately "5 feet 9 inches tall."

Orr claimed the creature ran across the highway and then "made a loop" toward his car, so he quickly drew his .357 Magnum revolver and ordered it to "halt" before firing a warning shot. When the creature did not comply, he shot at it several times. According to the airman, "one shot hit the creature in the neck and [it] leaned against the car." The creature managed to regain its footing, however, and ran into the woods, never to be seen again.

Orr told Truesdale that he got out to inspect the area of the hood where the creature had momentarily leaned. There he found traces of blood and scales, which he promptly collected with a napkin. He turned the evidence over to the police as proof. He also provided a drawing of the creature, which looked something like an alligator man with a long tail.

The airman stood behind his story, but Truesdale wasn't buying it. Not for a moment. First off, the sheriff could tell that the scales had come from a fish. The sheriff was no ichthyologist, but he'd been fishing enough times to recognize common scales. As well, the sheriff felt that the sketch Orr provided looked too much like the cartoonish "Lizard Man" creature depicted on a T-shirt being sold by a local vendor. The fact that Orr drove a camouflaged Toyota outfitted with fake machine guns didn't help his credibility either.

While Orr was at the station giving his statement, one of Truesdale's deputies inspected his outlandish vehicle. According to sheriff's department spokesman, Bill Moore, in an article that appeared in *The Item* newspaper on August 12, 1988: "When the deputy walked up to (Orr's) car, the gun was laying on the passenger seat." This was a problem since Orr had no permit for the gun. State law only allowed unregistered firearms to be carried if they were stored in the trunk or glove compartment of a car. Given this, Truesdale decided to charge Orr with a misdemeanor weapons offense which could net him up to one year in prison, a fine of $1,000, or both. He would also be charged with a misdemeanor count of filing a false report. It was the perfect way for the Lee County officials to send their own warning shot to the public. They simply would not tolerate random acts of gun firing or false reports concerning the Lizard Man.

Orr had been permitted to leave the sheriff's office that day, August 5, citing a doctor's appointment at the air base, but was later notified that a warrant had been issued for his arrest. By August 12 he had returned to the police station to admit that the entire story had been fabricated for the purpose of perpetuating the mystery. "He admitted it was hoax," Truesdale said. "[Orr] said he wanted to keep the legend of the Lizard Man alive."

Kenneth Orr, for all his good intentions, did generate more Lizard Man press, but not the kind he had intended. Chris Davis, on the other hand, had become something of a celebrity by now. He would show up at Bishopville's makeshift "Lizard Man Information Center" (located inside the Cottonland Restaurant) to sign autographs and answer questions. Later he was approached Joye Reis, a businesswoman from Sumter, who signed on as his manager. In no time she was charging for his appearances at local stores. In one appearance, he signed autographs at Myrtle Square Mall (in nearby Myrtle Beach) as Belk store employees sold T-shirts bearing the likeness of Davis and a drawing of a reptilian monster. According to an article that ran in the *Charlotte Observer* on August 14, 1988: "During the first hour of Davis' visit, store employees said they sold about 50 of the $13 T-shirts." Davis, who reportedly made more than $3,500 from autographs and T-shirt royalties, had mixed feelings about the whole thing. The money had been nice, but the ridicule and constant public attention didn't sit well with the shy teen. "I didn't want it to happen, but it did," Davis told a reporter from *The State* newspaper later that fall. His personal appearance schedule had also caused him to miss much of his high school's basketball season. It was a sport he loved and excelled at.

When asked if he would do it again (i.e., report his Lizard Man encounter), Davis simply replied: "No, it wasn't worth it. I couldn't do my job, and I couldn't play basketball."

On August 18, Davis finally had his chance to silence the naysayers. As promised, he was given a lie detector test administered by Sumter Police Captain Earl Berry at the Sumter City-County Law Enforcement Center. The test was paid for by Rick Welch, a former Sumter Police detective, and his partner, Randy Galloway, who comprised a firm called Southern Marketing. They had been working with Davis at some of his personal appearances when they saw some boys giving Davis a hard time. In an interview, which ran in *The State* newspaper on September 3, 1988, Galloway explained that they wanted to satisfy their own minds, as well as show the public that the young man was not lying. "Some in the media are saying he might be on drugs," he said, "so we wanted to set the record straight."

Some of the questions put forth to Davis included:

> Was the creature that attacked your car green and black?
>
> Were you drinking or smoking drugs?
>
> Were you really driving 35 mph when a creature jumped on your car?
>
> Did it occur immediately following your changing a flat tire?

Davis answered each question as Captain Berry evaluated his reactions. At the end of the test, Berry concluded that Davis had been truthful on all accounts. Of course, the effectiveness of polygraph testing is highly debated—so much so that the results are rarely accepted into court as evidence—but nonetheless it helped show the public what Sheriff Truesdale had believed all along… that something had truly scared Davis on that dark night in June.

Memorable photo of Chris Davis circa 1988
(courtesy of The Item*)*

The Butterbean Shed

For those who believed Davis' story, the question was not whether he had actually *seen* something that night, but rather *what* had he seen. Was it some sort of reptile-human hybrid, a man in a costume, or something else? The captivated public kept the questions coming, but for a few Bishopville locals, the answer was already clear. Though no one had admitted anything publicly, hushed conversations around town circulated a rumor that a local man had unwittingly given rise to the Bishopville Lizard Man when he chased a person fitting Davis' description from his property on Browntown Road in late June.

The man's name was Luscious "Brother" Elmore. He was a local farmer who specialized in a variety of Lima bean known in the South as a "butter bean." Over the years he had built a steady business selling to grocery stores and restaurant chains around the state, making him quite successful. Brother, as the locals called him, was a hard working man with a solid reputation, so it seemed ironic that he would end up at the center of a monster controversy.

As the story goes, Elmore was having a problem with air-conditioning units being stolen from his farm. The farm was located on Browntown Road about half a mile from Scape Ore Swamp and about half a mile from where Chris Davis said he had a flat tire. About 30 yards from the road onto the property sat a small building, known locally as the Butterbean Shed, where Brother and his employees dried and stored the butter beans. The shed was equipped with several air-conditioning window units to keep the workers cool during the hot summer. These were visible from the road and completely unguarded at night, making them a tempting target for unscrupulous thieves.

During my investigation, I spoke to a long-time resident and friend of Elmore who gave me his version of the story. The individual wishes to remain anonymous, so we'll call him "Steve." According to Steve, around the time of the Davis incident, Brother had purchased several new units for the shed. "He didn't want the new air-conditioners stolen so he decided to stay up there at night," Steve explained. "On the second night he heard a car stop out on Browntown Road." Thinking it might be a potential thief, he wrapped himself in a burlap sheet and bolted out of the shed.

In the moonlight, Elmore could see a car parked up the road and a person running towards it, so he began to give chase. Having popped up seemingly out of nowhere and draped in the sheet, he must have been a startling sight. The unknown person screamed, jumped into the car, and quickly drove away. "He had that sheet on and he was a big ol' tall guy, probably about six-three or six-four," Steve told me. "He scared off whoever it was in the dark."

The person had presumably been trying to steal one of the air-conditioners, since Elmore found one lying in the fennel grass between the car and the Butterbean Shed. Satisfied that he had prevented a theft, Elmore returned to the shed and slept there the rest of the night. Weeks later when he saw Davis' bizarre story in the newspaper, he wondered if, in fact, Davis had been the person he chased that night. It seemed highly coincidental.

Elmore didn't go public with his suspicions but did tell a few friends who began to circulate the story. It would be several years before he discussed the incident directly with Steve, however. "At the time, we all suspected it was Elmore, but we didn't know for sure," Steve told me. "I didn't tell anybody it was him until he told me several years after the thing died down. Elmore never said it was Davis, he just said 'I chased a guy who was out there with one of my air-conditioners.' I'm assuming it was Davis since Davis reported it happened right out where the shed's at. And it happened the same night as he chased somebody off."

Other Bishopville locals that knew Elmore told the same story, albeit with slightly different details (as happens when stories are passed around). In a 1998 interview with Alicia Lutz from the *College of Charleston Magazine*, local feed store owner Al Holland said that Elmore didn't bolt out of the shed that night but instead acted more stealthy. "He walks out to the road, which is lower than the yard, so he's up high, hiding in the dog fennel," Holland explained. "So he's standing there, he's looking down, when the kid turned around and screamed and took off."

Holland also provided a theory as to why Davis described the "creature" as having large red eyes. According to the article: "The way Holland tells it, Davis' taillights reflected in Elmore's glasses, causing an illusion of red glowing eyes; and the scrapes and scratches on his car were from the still-attached jack."

Steve saw the car damage himself. "To this day I don't know what happened to that boy's car," he admitted. "Maybe he took something and scratched it up himself."

I asked Truesdale for his opinion of the story. He said that, like some of the other locals, he heard about Elmore's alleged involvement in the days after Davis came forward. At the time, he could not rule it out, but he was never convinced that Elmore had been responsible. Later, Truesdale's intuition was seemingly confirmed. "Elmore said he did it at first, but when it got down to the nitty gritty, he said he didn't do it," Truesdale explained. "It

The now dilapidated Butterbean Shed pictured in 2012 (photo by Cindy Lee)

was probably a year or two after all this happened that the press came down on him and he denied every bit of it. Even his daughter denied it. She said he was at home those nights [when the incidents allegedly occurred]."

With such conflicting stories, we may never know the actual truth, especially since Elmore has passed away. But even if Elmore *did* surprise a would-be thief that night, as he originally claimed, there are several reasons why this does not automatically make him "the Lizard Man."

First, no one can really be sure that the person Elmore chased that night was Chris Davis, not even Elmore himself. It would certainly be coincidental that Davis reported having a flat near the Butterbean Shed on the very night Elmore chased away a thief, but *coincidence* is not *confirmation*.

Second, why would Davis make up the part about the creature jumping on his vehicle? In no account did Elmore say he jumped on the boy's car or chased him down the road at 35 miles per hour. If Davis had been frightened by a farmer he believed to be a large green humanoid with glowing red eyes, fine, but why make up the extra details? It just doesn't make sense.

Next, how do we explain the damage to the side mirror and roof of the car? If the car was damaged by the "still-attached jack," as Holland claims, then wouldn't the damage have been located closer to the fender or the door? Sure, Davis could have damaged the car himself to back up his monster story, but would a kid who worked for minimum wage at McDonald's want to bust up his own car?

Lastly, even if Davis mistook the farmer for a monster, it doesn't explain the other sightings. What about George Holloman and George Plyler, both of whom claimed to have seen a strange creature prior to Davis' alleged encounter? Was Elmore responsible for those too? The Davis sighting has become the most famous due to the frightening details, but it's certainly not the only sighting. And there would be more to follow, making it even more difficult to pin the entire phenomenon on the farmer. As we would soon learn, if Brother Elmore was *the one and only Lizard Man*, he would have had to be in two places at once!

Flying High Again

Sometime after Holloman and Plyler came forward with their stories, another man decided it was time to tell his own. Like the other men, he had allegedly seen something he could not explain but had thus far kept it quiet. Once the stories of the Lizard Man starting hitting the papers, however, his wife urged him to go public.

The man was a local cropduster by the name of Frank Mitchell. Over the years, Mitchell had built a steady clientele serving numerous farms in the area. He lived on a large expanse of property near Scape Ore Swamp, where he had a small airplane hanger and a private runway from which to operate. Each morning at daybreak he would gas up his plane, load it up with fertilizer, and take to the skies above the endless rows of corn and cotton.

I had the opportunity to interview the now retired cropduster during my visit to Bishopville. As Truesdale, Cindy, and I sat in

Mitchell's home drinking coffee, he went through the particulars of the incident. Unfortunately, he couldn't recall the exact date, only that it had occurred sometime before Chris Davis' sighting, perhaps a month prior.

On that morning, a dreary fog had settled in, so he was delayed in getting airborne. It was well after daybreak before the sun burned away the moisture, making it clear enough to fly. At that point, Mitchell taxied his plane to the runway and set off down the 3,000-foot grassy strip, which cut a straight path through the surrounding woods.

"I got down the runway, at the point I had to take off, and that's when this *thing* just walked across the runaway right in front of me," he recalled in his infectious Carolina accent. "But it didn't run; it kinda had a lope in its walk as it came across the runway and just looked at the airplane."

As the aircraft lifted into the air, Mitchell struggled to get a better look at the strange humanoid creature. "I couldn't tell if it

Mitchell's runway
(photo by the author)

was scaly or hairy, but it was a grayish-brown thing with a face like a monkey," he told us. "When I got in the air, I made as quick a turn as possible, but by the time I got turned around it was gone." The thing, whatever it was, had simply come out of the woods on one side of the airstrip, loped across the open grass, and disappeared into the tall trees on the other side. The sighting gave him a chill even from the air.

Mitchell went ahead with his aerial applications that day, wondering all the while what he could have possibly seen. Since this occurred before he had any knowledge of the Lizard Man or Chris Davis, it was something he could not quite understand. So he felt it would be best to keep the sighting to himself.

"I didn't say anything at the time," he explained. "I just didn't want my neighbors and my customers to think I'd completely gone crazy. I mean, they know I'm crazy by the way I fly," he chuckled, "but I didn't want 'em to think I'd gone completely loco."

When news of the Lizard Man hit the papers, he began to reconsider the incident, thinking that perhaps he had also seen the creature. He had certainly seen *something*, so he finally told his wife (who, up until then, had no knowledge of the incident). She urged him to tell the police. "When other people had seen probably the same thing, then I didn't care who knew it," he explained.

I asked Mitchell if he thought the creature was the so-called Lizard Man. He shook his head. "I don't know how that thing come up being called 'the Lizard Man,'" he said. "I didn't see a tail. What I saw was more like a tall monkey standing on its hind legs, walking just like we do except it had a lope in its walk."

I was beginning to see a pattern with the eyewitness descriptions. With the exception of the hoaxer, Kenneth Orr, thus far no one had mentioned seeing a tail. Not even Chris Davis. Mitchell agreed. "And the thing of it is, at the Butterbean Shed where Chris Davis was at, it was told that was a hoax and the guy who owned the shed was responsible for all of it. But that's baloney 'cause Brother [Elmore] was standin' in my yard the day I saw this thing. Brother was next to be sprayed, and him and another farmer were standin' in my yard talkin'!"

I glanced at Cindy. Without prompting, Mitchell had cast another shadow of doubt on the Elmore theory. I looked back at Mitchell. "So Elmore was there in your yard that morning?" I asked. Mitchell nodded. "He and [*name withheld*] were right there, so I know it wasn't him."

Mitchell offered us another cup of coffee before adding one more thought. "You know, I've had a lot of people come up and say 'Frank, go ahead and tell me the truth about this'... but that *is* the truth," he stated flatly. "I've never had a gun at my back tellin' me what to say and I've got no reason in the world to lie about it. I did have a reason to keep my mouth shut at first, 'cause those farmers pay me pretty good. But that's what happened."

After the interview, our conversation turned to music. Mitchell is also a guitar player and singer, and frequently performs around Bishopville. After learning of my own music background, he offered up one of his electric guitars, which I strapped on and strummed. He didn't know much in the way of rock songs, but we found some common ground with vintage country tunes. Having grown up in Texas, I'm familiar with many of the classic artists and songs, but Mitchell is a virtual jukebox. In no time he was singing a few Merle Haggard and Johnny Cash songs as I played rhythm. Truesdale suggested we try a song or two by one of his favorite artists, Kenny Rogers, but that's where things started to get a bit dodgy. Somewhere around the last chorus, which had us singing "you picked a fine time to leave me Lucille," I thought it might be best to quit while we were ahead. Cindy was still smiling, but I didn't want to push it.

After wrapping up the impromptu performance, Mitchell offered to take us down to see the runway. I was eager to see the very spot of his sighting, so we all piled into his SUV and headed across the property. We pulled up to a small airplane hanger several hundred yards away and stopped. To the left we could see a long, grassy strip of land that cut a wide path through massive trees. Cindy and I got out and walked toward it. We could see that there was enough woods on both sides of the airstrip to allow a large animal to roam virtually undetected. Mitchell's land is west of the main swampy bottoms but still lies within a reasonable proximity to the other sightings. The mental image of an airplane runway doesn't fit the typical scenario for a cryptid encounter, but after seeing this rural setting, I had little doubt that a sighting here was at least possible.

Looking to the far end, I could see where the clear-cut stopped and the trees resumed their hold on the land. A pilot would have to take off somewhere before that point if he were to have any chance of getting airborne. I imagined a large creature loping across that wide grassy area and what that must have looked like to Frank on the day he saw it. I could empathize with his frustration in not being able to turn around in time to get another look. Encounters like this always happen so quickly, leaving the bittersweet taste of mystery to linger forever.

After a few photo ops, we thanked Mitchell for his time and headed back to our cars. As we drove away, I thought more about what he had described. This was not a fantastical walking lizard, this seemed more like a man-ape. Could these people have seen a creature that in most instances would be called a "Bigfoot"? The

Cindy with Frank Mitchell
(photo by the author)

more witnesses reports that I came across, the more I started to think this might be the case. And, as I would eventually learn, I was not alone in this thought.

3. Lizardmania

The morning after our arrival in Bishopville, Truesdale took us to visit the South Carolina Cotton Museum. Located in the heart of downtown Bishopville, the museum is dedicated to preserving the legacy of cotton and rural life in the area by displaying a variety of farm and manufacturing equipment dating back nearly two centuries. Their exhibits don't include a Lizard Man section, but the museum's executive director, Janson Cox, still feels it's an important part of the local history. For that reason, the museum has become a central location for the preservation of all things Lizard Man.

"Some of the locals play down the Lizard Man stories," Cox told us, as we sipped coffee and thumbed through the museum's hefty collection of Lizard Man related news articles. "But it's an undeniable part of our local culture." And indeed, judging from the massive press coverage, both new and old, it would be hard to live in this area and not know of the Lizard Man.

After reading through some of the articles and discussing the purpose of our research trip, Cox led us to a back room, which serves as a storage archive. The modest space was filled with shelves and file drawers containing a variety of artifacts. As we began to look around, Cindy and I could see tangible evidence of the Lizard Man's impact on the town. One of the shelves was loaded with colorful T-shirts silkscreened in a variety of Lizard Man designs. Each shirt was carefully folded and sealed inside clear plastic. On another shelf we could see plaster casts of the famous three-toed footprint discovered in 1988. I examined the huge blocks, noting the curious and exaggerated track that was preserved within. Unlike many alleged Bigfoot/monster tracks, this footprint was well-formed and very clear. It looked too perfect and smooth to be from a real animal, and that's probably why wildlife officers had so readily

Track casting stored at the South Carolina Cotton Museum (photo by the author)

dismissed it. Nevertheless, it's part of the Lizard Man history so it still holds interest in that regard.

Moving on from the tracks, Cox showed us a large, green mask that looked something like a horse-faced dinosaur with a bad toupee. "That's the head of the old Lizard Man costume," he told us, explaining that the costume had been worn in the town's annual Cotton Festival Parade. In 1988, the Lizard Man usurped the former "Cotton King" as mascot. The long-running event was even renamed "The Lizard Man Festival" at one time to capitalize on the creature's fame. A second costume, which we also examined, looked more like a bright green version of Barney the dinosaur, the popular children's television character. At some point, the festival's organizers thought it would be better to make the creature look more family-friendly. I was partial to the old creepy one, but I could understand why it might come off as a bit frightening to young tots.

Next, Cox showed us the museum's copy of a 7-inch vinyl record that featured a song called "The Lizard Man" by Bishopville native Jim Nesbitt. Nesbitt was a moderately successful country artist who had scored a few hits on the Billboard charts back in the 1960s.

One of the Lizard Man costumes used in the Cotton Festival Parade (photo by the author)

During the summer of 1988, he was inspired to write and record "The Lizard Man," a song dedicated the mysterious monster of his hometown. The single sold well at the time, but as fate would have it, this would be his last studio recording.

Cox also pointed out an old wooden sign that had once hung on a certain infamous structure near Scape Ore Swamp. In faded white letters it read: "Elmore's Butterbean Shed." The director explained that no matter what was true or not true about the Davis incident, the shed was still an integral part of the Lizard Man mystery, and as such, he was fortunate to have the original sign. The building still stands near Browntown Road, but it has long-since collapsed into a pile of weathered wood, red bricks, and eroded cinderblock walls.

As we continued to sift through the artifacts—including the museum's own copy of the "Bishopville Lizardman" action figure and a collection DVDs from popular television shows featuring the Lizard Man—it was undeniably clear that the monster had made a huge impact on the area. But how had it gone from local lore to international fame? Certainly modern-day television shows about the Lizard Man—including *Animal X, Destination Truth, Fact or Faked: Paranormal Files, Lost Tapes, Mysteries at the Museum*, and *Weird or What?*—have helped keep the interest alive, but the Lee County legend was already famous long before these programs came along. As I dug more into the press archives, the answer would become obvious. From the very beginning, the mainstream media just couldn't resist the Lizard Man!

Feeding Frenzy

While sightings of regional Bigfoot creatures, alleged Chupacabras, lake monsters, and other mysterious beasts often make local headlines—sometimes even hanging around in the news for weeks—the media frenzy surrounding the Lizard Man case was staggering. Not since the early 1970s when a Bigfoot-like creature near the small town of Fouke, Arkansas, sank its claws into the headlines and inspired a movie sensation known as *The Legend of Boggy Creek* had a cryptid catapulted to international fame in such a short span of time.

Following the initial wave of regional newspaper coverage—including *The Item* (Sumter), *The State* (Columbia), *Lee County Observer* (Bishopville), and the *Independent-Mail* (Anderson)—a range of papers from neighboring states began to follow suit. Within a matter of days, outside reporters became as thick as the local reporters at any given time in Browntown.

When WCOS-FM leveled its million-dollar bounty on the creature, the story became even more famous. Soon the Associated Press and United Press International were circulating the stories. The Lizard Man's tale eventually wrapped itself half way around the world when the *Korean Times* wrote of the Waye's vehicle damage and the Davis incident. Soon after, the Lizard Man exploded onto the national television scene. I spoke to Skeet Woodham, who lived in Bishopville at the time. He remembered how the whirlwind of TV coverage got started.

"We had a screen printing shop back in those days," Woodham told me. "And what happened was we started making Lizard Man shirts and hats to sell to all the people coming into town. I had a hat made that said 'Lee County Lizard Patrol' with a red Lizard Man on it."

Woodham had worn the hat to the local Cottonland Restaurant. While he was there, another local informed him that a group of customers had been asking about the Lizard Man. When they saw Woodham's hat, they asked to be introduced. "It was a group of guys that were coming from Atlanta up the interstate going back to New York," Woodham explained. "They worked for NBC, as well as I can remember, because one of the guys told me 'if you'll give me a hat, I'll take it back and give it to Tom Brokaw and it'll be on national news.'"

Woodham did them one better and took the guys back to his shop where he gave them all complimentary hats and T-shirts. "Sure enough, a few days later Tom Brokaw made a comment about [the Lizard Man] on the national news," Woodham told me. At the time Brokaw was hosting the *NBC Nightly News*. "I can't remember what he said, but the Lee County Lizard Patrol hat was sitting right there on his desk during the show."

For the unsuspecting town, it seemed to be just the right combination of eerie intrigue and happenstance to start a media wildfire. Woodham agreed. "In my mind, if the hat and stuff had never gone to New York, I don't think it would have ever got out of the local region," he conjectured. "I think with the slow summer and those guys being in Bishopville and hearing all those stories, that's what kickstarted it."

Soon, more television shows were calling. *PM Magazine*, NBC's *Today* show, the Fox TV network, CNN, and *Unsolved Mysteries* all phoned the Sheriff's Office trying to get a piece of the action. Not long after, the Lizard Man story made a jump back to the airwaves when popular radio host Paul Harvey brought up the case on his ABC Network syndicated show. This brought Bishopville to the attention of millions across more than 1,200 radio stations and 400 Armed Forces networks.

Rendition of the Lizard Man from The Charlotte Observer, August 1988 (courtesy of Loyd Dillon)

But it didn't stop there. On July 23, 1988, journalists from *People* magazine made their way to Bishopville. According an article in *The Charlotte Observer*, "People magazine reporter Linda Kramer talked with residents as her photographer snapped a picture of an inflatable Lizard Man behind a tree." The journalists, it seemed, were willing to do anything to get their own scoop on the story, even if that meant braving the hot South Carolina summer.

Two days after *People* dropped by, Dan Rather from the *CBS Evening News* interviewed Sheriff Truesdale. The interview was part of a special edition report that also featured Chris Davis as he recounted his harrowing Lizard Man encounter, along with other locals who conjectured about the reality of the phenomenon. The segment closed out with audio from a song called "Lizard Man."

The *Oprah Winfrey Show* was next to consider the Lizard Man story. Representatives from the show reportedly reached out to Chris Davis, but an on-air interview never materialized. Television coverage finally reached its peak on July 29 when *Good Morning America* broadcast live from Browntown with Truesdale as their guest. "This thing was all over the television," Truesdale quipped during one of our conversations. There was just something about this small-town monster that fascinated the public.

By August, the coverage had abated, but the frenzy was rejuvenated when CNN provided an update on the happenings. Magazines such as *Omni* and *Time* even mentioned the incidents. *Time* concluded: "The monster may be the biggest thing to hit Bishopville." Given the press circus, there was little doubt.

The bridge at Scape Ore Swamp circa 1988
The graffiti reads: "Carol & Janet Came To See the Lizard Man"
(photo by Rachael Bradbury)

With all the attention, the Lee County Sheriff's Office had become the prime target for press inquiries and crazy phone calls. "We were getting 200 calls a day from around the world," Truesdale told me. "England called four times one morning to ask about motels near Browntown."

To deal with the influx of calls, Truesdale appointed Billy Moore, the records clerk at the Lee County jail, full-time to the task. This took some of the pressure off the sheriff and his deputies, but still they were having to spend much of their personal time in the evenings catching up on casework. And then there was the problem of personal privacy. People were calling Truesdale at his home at all hours. In one memorable instance, a radio disc jockey called at 6:00 a.m. "He asked me what I was doing," Truesdale told reporters from *The Item* on July 31, 1988. "I just asked him what he *thought* I was doing that time of the morning. You try to be nice to them because they do have a right to know what is going on."

Truesdale was careful to maintain a good public image with the press. "We don't want to be played up like some kind of Mayberry—

A newsprint ad exploiting the Lizard Man craze circa 1988

like a bunch of country hicks," he explained. "One person from out of town said that a lot of people are laughing at us, but we should be laughing at all of them. They are the ones who are coming as far away as the West Coast."

Truesdale certainly had a right to be frustrated by the situation, but even so he realized it was something he could no longer control. No one could have ever predicted such a reaction to the stories. "Personally, I have had enough of the Lizard Man," he said at the time. "But this thing is no longer our news. It is the world's news and we simply can't contain it."

DAYS OF OUR LIZARD

During the frenzy of "Lizardmania," as one newspaper dubbed it, Bishopville was constantly flooded with throngs of curious sightseers, gawkers, and monster hunters. Given this, it was only natural that some enterprising entrepreneurs took advantage of the opportunity. Local businesses began playing up the Lizard Man theme by putting out signs or blow-up dinosaurs in front of their establishments to draw in more customers. Many of these businesses also sold their own Lizard Man souvenirs such as T-shirts, hats, and buttons.

Alva Kelley, the owner of an Exxon station in town, told reporters from the *Lee County Observer* on August 3, 1988 that he was selling at least a dozen shirts a day to tourists. The shirts depicted a rather goofy drawing of a "lizard man" with the words "Wanted: Lizard Man Alive—1,000,000 Reward," but that didn't seem to matter. People were simply obsessed with the Lizard Man and would buy any sort of memorabilia they could take with them. "We have people stopping by all the time to buy shirts and ask directions to Browntown," he said in the article.

Some of the entrepreneurs took their operations right to the main hub of activity. They would park on Browntown Road and sell their wares right out of their vehicle. In my conversation with Skeet Woodham, he described the scene well. "From Browntown

Front window of Kelley's Exxon during the summer of 1988 (photo by Rachael Bradbury)

Road on down to the Interstate, none of the stuff that's there now was built at the time. You could see cars three deep on each side of the road with people selling shirts. A bunch of guys out of Charlotte [North Carolina] even came down in great big ol' trucks."

The guys from Charlotte were not the only outsiders to try their hand at Lizard Man marketing. Gail Gaddy, the owner of a tanning salon in Columbia some 50 miles away, brought in a batch of her own merchandise. According to *The Herald Independent* on Aug. 11, 1988, "Gail had heard that a festival was going to be held in the area and had taken along some specially-made Lizard Man T-shirts and hats and other items. Instead, she found the road through Scape Ore Swamp lined with cars for miles. She and her companions arrived about 6 p.m. and promptly set up shop selling the Lizard Man novelties. By the time she and her companion left, a little after midnight, she had sold about 200 of them."

Other entrepreneurs put aside their real jobs to take up the Lizard Man business full time. Randy Galloway, an insurance salesman from nearby Sumter, was one such person. According to an article published in the *Anderson Independent-Mail* on July 31,

1988, Galloway had given up his insurance business to set up a roadside stand next to the local Cottonland Restaurant. He began selling T-shirts and bumper stickers with such success that he planned to add 11x14 inch photograph prints of the famous three-toed tracks. "This is a full-time job," he told reporters.

The amount of merchandise being sold locally was astounding, but the demand went beyond Bishopville. Skeet Woodham remembers exporting boxes of shirts to Europe on a weekly basis! "What made it so cool is that people from all over the world were calling and wanting stuff. Then their friends wanted stuff, and it just ballooned from there," he told me. "A lot of the guys over at Shaw field [air force base] were going back and forth over to Germany that summer. We were making shirts for them and they were taking them overseas, so we were just shipping shirts and hats everywhere."

On Sunday August 21, Lizardmania finally reached a high point (or low point, depending on how you look at this) when a well-known soap actor took part in Bishopville's Cotton Festival. Drake Hogestyn, who at the time portrayed police Commissioner Roman Brady on the popular NBC soap opera *Days of Our Lives*, joined the festival to help raise money for the town. As part of his appearance, he interviewed Chris Davis. The two appeared on stage together as "Commissioner Roman Brady" quizzed the teenager about his

Vintage Lizard Man T-shirt designs
(photos by the author)

encounter with the Lizard Man. Following the interview, Hogestyn danced and performed skits with audience members before he and Davis signed autographs.

This event, while an overwhelming success with the crowd, exemplified what the Lee County Lizard Man had become in just a few scant weeks. What started as a truly curious and anomalous phenomenon had morphed into a spectacle of pop culture and merchandising. The quick transformation—and sheer amount of press coverage—was remarkable, but it completely overshadowed the question of whether there was something biological (whether known or unknown) behind the incidents. The Sheriff's Office was still taking reports and attempted hoaxes seriously, but by now most people in the area viewed the creature as nothing more than fancy folklore. The media lost interest as well. There were some journalists who had always treated the subject with tongue-in-check, but by late August, the majority of newspaper writers took the humorous slant with their articles or ignored the Lizard Man altogether. It was understandable, given the ridiculousness of dancing soap opera stars and autograph sessions, but still questions remained and sightings continued.

Reptile Dysfunction

On August 26, 1988, Colonel Mason Phillips (name changed), a high-ranking official in the Army Corps of Engineers, was driving along McDuffy Road in the Ashwood section of Lee County. He and his wife had attended a wedding rehearsal that evening and were returning home when they neared a body of water called Ashwood Lake (approximately eight miles from Browntown). As Phillips rounded a curve, he saw a large, brownish creature emerge from the shadows and run across the road. It moved at a high rate of speed on two legs. In an article published on August 31 in the *Lee County Observer*, the creature was described as being "about eight feet tall with a tail that did not quite reach the ground."

His wife was dozing off at the time, so he called out to wake her. She raised her head in time to see the creature as it veered off the road and disappeared into a thick cluster of trees. The Colonel slowed the car as they tried to get a better look but did not stop. They were too shaken to investigate any further.

Colonel Phillips had never put any stock in the Lizard Man stories, but the incident made him change his mind. As such, he decided to report the sighting to the Lee County Sheriff's Office, although he requested his name be withheld from the public because of his high-ranking government position.[4] Truesdale took the report and felt that it was credible enough to investigate. The following day deputies were dispatched to the location where they searched for tracks or other signs of a large animal. Unfortunately, they found nothing. "Because of the rain we got most of the day Sunday, we were not able to identify any tracks," Truesdale told reporters from the *Lee County Observer*.

Truesdale had Phillips submit a written report of his observations and fully intended to have the story verified by polygraph examination, although the test never took place. A hypnosis session was also proposed—in case it could bring out additional details—but this never came to fruition either. Truesdale told the news: "Even though this person is not a run of the mill man, we plan to treat him as we have treated every other person who had reported seeing this creature."

The location of Ashwood Lake is a good distance from the sightings in Browntown, yet it's still within close proximity of Scape Ore Swamp and the location of Frank Mitchell's sighting. Ashwood is known as a stretch of bottomland where strange things have been said to occur. According to Joye Reis in the September 20, 1990, edition of the *Observer-Times*: "There have been strange noises heard in the swamp for years. Animals have been found pulled apart. And nobody's ever explained the woman whose head was twisted off at Ashwood Pond near the swamp, 15 years ago."[5]

---

4 His name has since been made public, but I chose to observe his original request and keep his real name off the record here.

5 I tried to find out more about this intriguing head twisting incident but could not find any further info.

The creature described by Colonel Phillips

Compelling as the Colonel's sighting was, it only deepened the mystery. One reason was because the Colonel reported the creature as having a "tail." Although many artistic renditions of the so-called Lizard Man in the media and on merchandise depicted the creature as having a tail, amazingly, no one prior to the Colonel had ever reported seeing such an appendage. His description varied from all the other witnesses' descriptions, which made it all the more confusing. Had he been influenced by the pop culture glamorization, or had the other witnesses missed this detail? It was a question that cryptozoology researchers began to debate.

The first such researcher to comment on the happenings was Eric Beckjord. Beckjord was a rather controversial figure who founded an organization known as The Cryptozoology Museum Project located in Malibu, California. After catching the Lizard Man story on television, Beckjord wasted no time in tracking down Chris Davis for an interview. As a result of the interview, and taking into consideration the other circumstantial evidence such as the tracks, car damage, and other sightings, Beckjord—in what might be considered one of his more astute moments—concluded that it was not a "lizard man" that was haunting Bishopville, but a type of Bigfoot. In an interview with reporters from *The Item*, Beckjord stated that the creature was most likely a "skunk ape," which is a southern version of Bigfoot said to live primarily in the southeastern portion of the United States. "You can call it anything you want, but it's a Bigfoot," Beckjord concluded.

One problem with this statement is that almost all Bigfoot witnesses describe the creatures as being covered in hair. Skunk apes in particular are often said to have longer matted fur that sometimes reaches four inches in length. Beckjord, however, accounts for this. In an official press release from The Cryptozoology Museum Project (7/27/88), he theorized why witnesses had described scaly skin instead of hair. "In such cases, the mud from the swamp may have dried and cracked, giving [the creature] a scaly appearance," he wrote. If the creature that attacked Davis was a Bigfoot, Beckjord felt it had been attracted to the "load of fishburgers in the car." Since Davis had been bringing home a sack of McDonald's Filet

O' Fish sandwiches, it seemed logical since Bigfoot witnesses have reported seeing the creatures trying to catch fish by hand in rivers or streams.

A second cryptozoology researcher, also drawn in by the news coverage, made his way from Ohio to Lee County. Although he was never identified by name, his opinion was given anonymously in an article printed in *The Item* newspaper on January 1, 1989. The individual had examined the trackway in Browntown, but concluded that they were "probably made by a horse with fake claws affixed to the hooves." However, he did feel that Davis' encounter had merit since "his account of the beast was typical of people describing their encounters with Bigfoots."

A sasquatch or skunk ape may seem like an equally far-fetched explanation to some, however, it may offer a more plausible reality than a reptile-human hybrid. There are those who believe "reptoids" walk in our midst (we will discuss this in more detail later), but in regard to the sightings by George Holloman, George Plyler, Frank Mitchell, and others, an ape-like creature could not be ruled out.

Creatures Among Us

One reason for the rampant sensationalism surrounding the Lizard Man may be due to our communal fascination for such a creature. Various concepts of "reptile men" or "fish men"—whether perceived as actual men or as gods—can be found throughout human history. They are woven into the very fabric of our mythology, religion, and folklore.

One such example dates back to ancient Greece, in which Cecrops, the mythical king of Athens, was depicted as having the upper body of a man and the lower body of a serpent. Another example can be found in Chinese mythology where Fu Xi—a cultural hero and inventor of writing, fishing, and trapping—is also said to be a hybrid of man and serpent. In ancient Egypt, the god Sobek was the deification of crocodiles, an animal that co-existed with people along the Nile River. As such, Sobek was often depicted as a man with the head of a crocodile. The Nāga were reptilian beings from Hindu mythology, said to dwell underground and interact with human beings on the surface. Even the talking serpent who tempts Eve in the Christian creation narrative is yet another example. This serpent, called *nahash,* was occasionally depicted as having legs and is most often associated with the character of Satan. Still more examples can be found in Aztec, Iranian, and Islamic mythos, just to name a few.

The presence of these creatures can also be traced to various regional folklore. Tales from Portugal and Galicia (Spain) include a supernatural woman known as *moura encantada.* Her character varies in description from story to story, but in some cases she is represented as a shapeshifter who takes the form of snake with wings or can appear as half-woman, half-animal.

Native Americans, as well, often told of their encounters with reptilian-like beings who they believed shared the world. The Hopi Indians, for example, centered much of their religion around a people they called their "Snake Brothers." These reptilians were said to have played an important role in their ancestral development.

In modern times, various incarnations of reptilian or amphibian humanoids permeate the realms of cryptozoology, alien encounters, conspiracy theories, books, comic books, games, and movies. Some of these are well known fictional characters, such as the *Creature from the Black Lagoon* and *Swamp Thing,* while some are nameless horrors originating with UFO reports and strange conspiracy scenarios that involve intelligent reptiles disguised as humans. But no matter what flavor these monsters come in, the fascination is always there.

Given this preoccupation, it is not surprising that the Bishopville Lizard Man sightings gripped the public as they did. Despite contradictions in the creature's description, this was the "Lizard Man," and to a certain extent, those two words influenced popular perception of the monster more than anything else. As Truesdale once said, "if the Lizard Man hadn't come along, they would've had to invent him."

There seems to be no doubt about that. Creatures such as this

Lizard Man

The fascinating concept of a reptilian humanoid

have existed in our society for many years, but none have been able to crossover into the "real world" with quite the same impact as the Lizard Man of Lee County. The creature presented one of the most intriguing cryptozoology cases, while at the same time providing the media with a "monster" they could easily personify and sell to an audience that was primed and ready. After all, it was a public that had grown up with monster movies. And whether they knew it or not, it was pop culture that helped pave the way for Lizardmania.

Pop culture is often driven by movies, and reptilian and/or amphibian humanoids have been sloshing their way through horror films dating back to at least the 1950s. The best and most famous example can be found in the *Creature from the Black Lagoon*, released by Universal International Pictures in 1954. The Creature, as the character is commonly called, was the last of the movie monster icons to come out of Universal Studios. Preceded by Dracula, Frankenstein, The Mummy, and The Wolfman, The Creature capitalized on Universal's supreme talent for creating memorable movies with truly iconic horrors. As such, the Creature secured its place as the seminal swamp monster, instantly recognizable by horror movie buffs and pop culture devotees alike.

Taking a brief detour to recap the *Creature from the Black Lagoon* storyline, we find that the film draws many parallels to cryptozoology, including several similarities to the Lizard Man case. First and foremost, both the film and cryptozoology rely heavily on the appealing notion that there are undiscovered creatures still lurking in the world's remote corners. In fact, the very inspiration for *Creature from the Black Lagoon* came from a cryptid that was said to exist in South America. In a 1995 *Starlog* magazine interview with Tom Weaver, *Creature from the Black Lagoon* producer William Alland told how the idea came about:

> During the making of *Citizen Kane*, I had dinner one evening with Orson Welles and his girlfriend at the time, Dolores Del Rio, and a tremendously famous Mexican cinematographer named Gabriel Figueroa. In idle conversation, Figueroa told the story about this creature that lives up the Amazon who is half-man, half-fish. Once

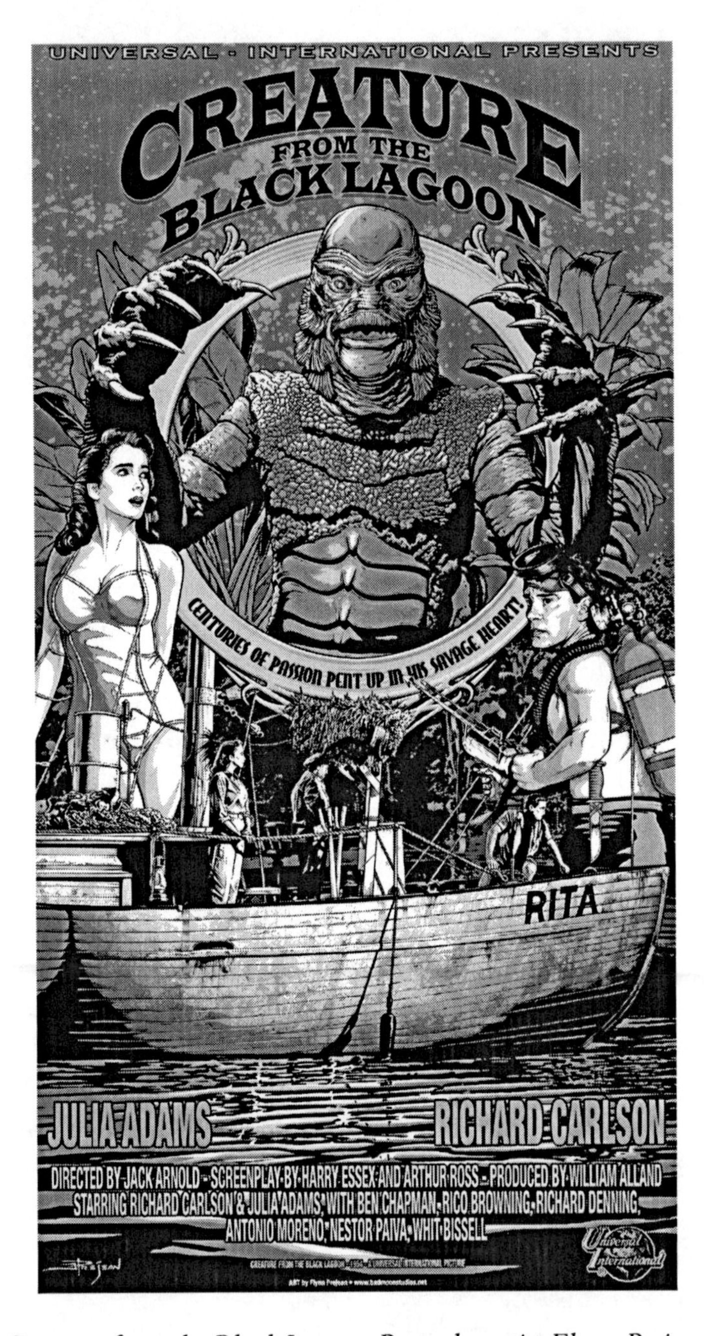

Creature from the Black Lagoon Poster by artist Flynn Prejean (courtesy of Bad Moon Studios—badmoonstudios.net)

> a year he comes up and claims a maiden, and after that, he leaves, and the village is then safe for another year. We just looked at him. He said, "You people think I'm joking, don't you?" and he then insisted that this was absolutely true, that he could produce photos! For about five minutes there, he held forth about how this was not a myth, that there really was such a creature, that the Amazonian people talked about him all the time, etc. So there it was, in my brain.[6]

As with the Lizard Man case, the film plays out like a modern day cryptozoology hunt in which individuals brave an inhospitable jungle in the hopes of uncovering evidence of a previously unknown animal. At the outset, Dr. Carl Maia (Antonio Moreno) discovers the fossilized forearm of some huge, clawed creature while doing research along the Amazon River. Anticipating the importance of this find, Maia heads back to civilization to enlist the help of Dr. David Reed (Richard Carlson) and Dr. Mark Williams (Richard Denning), who, along with their lovely assistant, Kay Lawrence (Julie Adams), return to the Amazon so that they can help Maia extract the fossil for the glory of science and, of course, monetary gain.

When the team arrives at Maia's encampment, however, they are horrified to discover that something has killed all of his native assistants. Figuring it must have been a jaguar, the doctors go ahead with their digging, but can't find the rest of the remains. Reed can only assume that the fossil must have washed down the river to an area known as the Black Lagoon. The team then moves downstream despite warnings from their boat captain, who tells them that many people have gone missing in the lagoon over the years.

Not long after entering the Black Lagoon, the team begins to suspect that something may be stalking them from under the water.

---

6 I did an extensive search to find more information about this alleged creature, but could not find anything specific to the Amazon. The only reference I could find to a half-man, half-fish in South America, was in Mark A. Hall's book, *Lizardmen*. He tells of a similar creature that was said to live along a river within the Chaco region of Paraguay. However, this is a considerable distance from the Amazonian region of Brazil.

When Miss Lawrence has a close encounter with something just under the surface, the men dive below where they get a glimpse of some large, aquatic creature that resembles a half-man, half-fish (played by Ben Chapman on land and Ricou Browning in water). When they return to the boat, the captain tells them of an old legend about a "gill man" said to inhabit the lagoon. Reed attempts to photograph the Creature while diving, but naturally, the images come up blank.

The action then ramps up as the Creature gets more aggressive. Eventually, he is able to kidnap Miss Lawrence, who is (understandably) the object of his affections. Reed and Williams argue about whether or not to kill the creature before it finally kills Williams, thus sealing its own fate. The Creature is then poisoned and subsequently shot in a series of tense confrontations as he tries to keep Lawrence in his clutches. In the end, the Creature slinks back into the black water, severely wounded and presumably doomed to die.

The fact that the creature turns out to be so human-like contributes to the movie's success. The pathos infused in the "gill man," like in so many of the other iconic Universal monsters, helps audiences identify with the character and ultimately feel the pain of his tragic confrontation with humans. This theme is also reflected in the Lizard Man case. If it's a human-like creature that walks on two legs, then it must have some level of higher intelligence. It is a thought is both tantalizing and sobering.

As well, the fact that Reed's photographs do not end up showing a clear image of the creature is also very reminiscent of cryptozoology. (No doubt everyone is familiar with the horde of blurry Bigfoot photos that plague the internet.) But unlike most cryptids, including the Bishopville monster, the Creature of the movie is far less elusive and eventually shows itself to the scientists. This is the payoff for audiences who are eager to see its face. The producers knew that the Creature's design could make or break the movie, so they strove to create something that would be worthy of their horror icon lineage. The result was a Creature that not only epitomized the modern day swamp monster, but ultimately imprinted its image in the mind of the public.

BEYOND THE BLACK LAGOON

Creature from the Black Lagoon is just one of many films that explores the reptilian/amphibian hybrid theme. After Universal produced two sequels, *Revenge of the Creature* in 1955 and *The Creature Walks Among Us* in 1956, other studios began to create their own spin-offs. The next to emerge was the far-fetched, but rather amusing, *The Alligator People*, released by 20th Century Fox in 1959. The movie tells the story of Joyce Webster (Beverly Garland) who goes to the Louisiana bayou country in search of her missing husband, Paul (Richard Crane), only to find that he's been turned into a sort of alligator-man mutant. In this case, science is responsible for the horrifying transformation— Paul was given a serum that was supposed to help him regenerate his limbs after suffering a horrible plane crash. The limbs apparently grow back but not without consequence, as he is slowly turned into a reptile. The result is a man with a scale-covered body and an alligator head.

While the movie is fairly entertaining—partly due to Lon Chaney Jr.'s performance as a cantankerous old Cajun—it is thoroughly immersed in a B-movie miasma despite the upscale pedigree of 20th Century Fox. The alligator-man comes off as extremely comical and completely unrealistic, placing it leagues below Universal's seminal Creature. However, interestingly, the alligator-headed creature resembles some depictions of the Bishopville Lizard Man that surfaced in the papers during Lizardmania!

Following close on the tail of *The Alligator People* came several other creature features that offered monsters of a reptile-human sort. These included *The Monster of Piedras Blancas* (1959), *The Hideous Sun Demon* (1959), *The Horror of Beach Party* (1964), and *Curse of the Swamp Creature* (1966). All of these were independent, low-budget productions with slipshod monster costumes that nonetheless carried on the tradition of swamp monsters in their own way.

Hammer Studios from England, famous for its Technicolor versions of *Frankenstein* and *Dracula*, released *The Reptile* in 1966. This film explored the skin-crawling combination of snake and human with the main protagonist being a "snake woman." This same theme was also explored by *Night of the Cobra Woman* in 1972 and *Sssssss* in 1973.

A slew of other films throughout the seventies continued to mine the public's fascination for reptilian/amphibian hybrids with releases such as *Graveyard of Horror* (1971), *Zaat* (1971), *Track of the Moonbeast* (1976), *When the Screaming Stops* (1976), and *Screamers* (1979). Most of these obscure films take liberties with the theme—such as having the moon trigger a transformation to a hideous reptilian in *Track of the Moonbeast*—but none go as far as *Zaat* (a.k.a. *Bloodwaters of Dr. Z*), which serves up a walking mutant catfish as its monster!

Lizard-like men also invaded the television screen in the form of "sleestaks." These scaly, green, bug-eyed humanoids were part of the popular Saturday morning show, *Land of the Lost*, which aired from 1974 to 1976. Their iconic look and shuffling walk made them appealing to young audiences, who embraced them just as their parents had embraced the *Creature from the Black Lagoon* years before.

In the same decade as the Lizard Man reports, B-movie mogul Roger Corman released *Humanoids from the Deep* (1980), which refreshed the image of an aquatic muck monster in the minds of the public. The movie was often played on early cable television and was available on VHS, so it was something that would have been somewhat familiar to people during the wave of Lizard Man sightings. Another similarly-themed movie, *Bog*, was released in 1983, although it's not as well known as *Humanoids from the Deep*. But again, it contained a slippery mutant monster in the tradition of the *Creature from the Black Lagoon*.

Comic books have long been proponents of the swamp monster, having produced Creature knock-offs in horror titles, super villains that were half-man, half-lizard (The Lizard of *Spider Man*, for example), and a character from DC Comics called Swamp Thing.

Promotional poster for Zaat (1971)
(Horizon Films)

In response to the overwhelming success of the comic book, a *Swamp Thing* movie was released in 1982. The movie, like the comic book, featured a half-plant, half-man antihero created by the fusion of man and swamp-matter. Interestingly, the movie was filmed on location in Charleston, South Carolina, just a two-hour drive from Bishopville!

Many more "swamp monster" movies have been released since the 1980s, but the point here is that the films preceding Lizardmania must surely have played a part in influencing the public's wild reaction to the Lizard Man sightings, as well as their perception of the actual creature. Yes, these movies were fiction, but to the general population it was the only reference they had for what a "Lizard Man" must look like. It was their very own, real-life Creature from the Black Lagoon; a monster not from the far-away Amazon, but from an overgrown and mysterious place known as Scape Ore Swamp.

4. Scape Ore Swamp

Truesdale held his gun high in the air trying not to get it wet. He was up to his chest in loose, watery soil, which seemed a lot like quicksand. He had always heard that there was quicksand in Scape Ore Swamp, but until then he was not sure he believed it.

Just moments before, he had been chasing a lanky moonshiner through the thick woods. Truesdale was a county police officer at the time, so part of his job was to locate and shut down illegal liquor stills. That morning he and another officer had crept up on one such operation hidden deep within the swamp.

"We knew the still was about ready to run, so me and another deputy went up there that morning to catch 'em," Truesdale told Cindy and me. "When we went in, we saw a man sittin' up there. He was real tall, about six-foot-four. And of course when they see the law comin', they run. Well, I had a good reputation of runnin' down people so I went after him."

It was no easy task to run through the terrain of tangled vines and sloshy ground, but Truesdale was determined not to let the man get away. He gave chase, moving as quickly as he could, holding his gun in one hand and swiping away low-hanging branches with the other. After a few hundred yards, the man took a sharp turn and headed for a clump of leafy water oaks. He was sprinting ahead of Truesdale at a good pace when all of a sudden the moonshiner plunged downward as if the earth had literally swallowed him up.

"There was this strip of water that looked like a little stream, but the water didn't flow, it was just still," explained Truesdale. The moonshiner had stepped into it and immediately sunk. "He hit that and went down. It came up to his chest, it was so deep."

The moonshiner used his momentum to propel himself forward as he grasped for the opposite bank of solid ground. Within a few seconds he managed to escape. Truesdale, unable to stop his own forward motion, plunged into the liquefied soil after him. It quickly gurgled up to his shoulders. Like the moonshiner, Truesdale hit the sink hole with force so there was just enough forward motion to keep him moving towards the other side. He wasn't completely stuck, but the situation was not good.

As the young deputy struggled, thoughts of drowning in quicksand flashed through his mind. All the debate about whether such a thing existed in Scape Ore was instantly moot. This might not have been the exotic setting of an Indiana Jones movie, but Southern swamps can hold their own when it comes to nasty surprises.

Truesdale, using every ounce of his momentum, managed to keep moving forward until he finally reached dry ground. Within a few more seconds, he lurched from the bog, shaken and muddy.

Truesdale was fortunate to have made it out alive, but he didn't have time to count his blessings. The moonshiner had been shaken up by his own pitfall, but his lanky legs kept running. The young policeman, more determined than ever, shook off the mud and resumed the chase through the swamp. A short time later, he was able to close the gap and bring the criminal down. Truesdale was soaked, dirty, and completely out of breath by the time he slapped on the cuffs, but he didn't care. It was all in a day's work for a Lee County deputy.

Truesdale finished up the story and shook his head. "You know, I think the reason I was able to get out was because of the speed I was going… it just got me on out of there. If you just stepped into it, you'd probably just stay there and that'd be it."

Scape Ore Swamp was indeed a primitive place full of remote hideaways and imminent danger. For this reason, the moonshiners and other criminals sought the protection of its dark, secluded depths where only the hardiest of law officials or hunters would be willing to venture. Aside from sinkholes and quicksand, there were thorny brambles and boggy bottoms to deal with. Worse, a host of deadly denizens slithered and lurked about, including snakes, spiders, cougars, and even bears. Any and all of these dangers were more than willing to lend a grisly hand of fate to anyone who dared to challenge the swamp's authority or suffer the misfortune of getting lost. And if that wasn't enough, there was, possibly, an even more nefarious denizen that called Scape Ore Swamp its home. So far it hadn't killed anyone—at least that anybody knew of—but that didn't mean it would not do so in the future.

The Lizard Man was certainly a concern, but the area's long and notorious history didn't start with sightings of this mysterious creature. The swamp had always been associated with ghosts and other strange phenomenon. It had also, supposedly, claimed a few lives of its own in the past. Even the swamp's name was the stuff of legends, having been changed several times over the years due to controversy. Scape Ore Swamp may be famous for its connection to the Lizard Man, but it has a lurid past all its own.

Home of the Monster

To get a better idea how the swamp fits into the landscape, it's important to first paint an accurate picture of the little town that shares its notoriety. Bishopville is known throughout the cryptozoology world as the "home of the Lizard Man," but unless you've visited there, you may not have a true mental image of just what the town is like or where it sits in relation to the swamp. Many times we concentrate so narrowly on the sightings of these "monsters" that the larger picture is ignored. Placing these creature sightings into the greater fold of the surrounding landscapes and hardscapes always lends a better perspective on whether these things might actually be real or simply urban legends.

Taking a closer look at Bishopville, we find a fairly typical small Southern town. According to the 2010 census, its population was roughly 3,500 residents, living within a total area of 2.4 square miles. Families, many of whom have lived in the area for generations, are dispersed among rustic side streets dotted with an assortment of houses that still reflect classic antebellum architecture. Large porches wrap the front of white-board houses, inviting one to sit a spell and enjoy a cold glass of sweet tea on a hot summer day. The town is not affluent by any means but nonetheless retains a sense of South Carolina charm that can only be found among the stately live oaks and costal palmettos for which the state is known.

Bishopville sits within reach of larger cities, such as Florence and Sumter, yet it still remains isolated along the old U.S. Route 15. The closest major thoroughfare is Interstate 20, but several miles of roadway and a swath of thick Carolina countryside manage to shield the town from its encroaching hum.

Welcome to Bishopville
(photo by the author)

Three miles into US 15, the road slows pace to become Main Street. Here we find Bishopville's downtown, located at the center point of the town's circular perimeter. The downtown is made up of several blocks of brick buildings, not unlike other small towns across America. Here the traditional flavors of Southern architecture blend with modern additions to provide facilities for the *Lee County Observer*, the public library, and the Cotton Museum, along with an assortment of merchant stores, new and old. There's even a Piggly Wiggly grocery store nearby, a veritable symbol of early Southern enterprise. The supermarket chain, which got its start in 1916, was prominent throughout the South and eventually reached markets in the central United States. I remember having a Piggly Wiggly near my home in Texas when I was young, but haven't seen one since it went out of business in the late 1970s. I nearly jumped over the curb trying to veer off into the Piggly parking lot, as fond memories of the oddly named store came rushing back. Cindy and I felt compelled to make a pit stop in search of Piggly Wiggly shirts to bring back as souvenirs.

But alas, as Main Street continues, the charm of Piggly Wiggly gives way to more distinguished landmarks. On one corner, a life-size bronze statue catches the eye. Long before the Lizard Man, Bishopville was best known as the birthplace of college football legend, and 1945 Heisman Trophy winner, Felix "Doc" Blanchard. Blanchard was the first junior classman to ever win the Heisman Trophy, and the first-ever football player to win the James E. Sullivan Award, both in 1945. As such, his legacy has been proudly commemorated by this large statue.

One might expect to also see evidence of the Lizard Man's legacy along Main Street, yet we saw nothing. There were no statues, murals, or even business logos that incorporated the creature's image. The only place we could find that outwardly celebrated the Lizard Man was a restaurant called Harry & Harry Too. This family-owned restaurant, located a mile from downtown, has an impressive wrought-iron sign that not only incorporates the creature's image but announces it as "Home of The Lizard Man."

Curious to see the "Home of The Lizard Man," Cindy and I accompanied Truesdale for lunch one afternoon. Inside the restaurant we found a few mementos on display, including a T-shirt and one of the original three-toed track castings. The menu itself featured a specialty sandwich called The Lizard Man, so we promptly ordered a round. The sandwich—consisting of chicken tenders, sautéed mushrooms, onions, and melted provolone stuffed into a hoagie roll—turned out to be rather tasty. During the meal, we also had the pleasure of speaking with the owners, who shared our fascination for the Lizard Man phenomenon. We spoke at length about the creature's significance to the local culture and the intrigue of its on-going mystery.

Street sign for the Harry & Harry Too restaurant (photo by the author)

Bishopville, I learned, is also home to the renown Pearl Fryar Topiary Garden. This amazing landscape of carefully sculptured trees and shrubs, which lies on the southern outskirts of the town, was created by resident Pearl Fryar on the three-acre property surrounding his home. Fryer and his garden have been featured on various television shows, and most recently in the critically acclaimed documentary, *A Man Named Pearl*. Cindy and I had the opportunity to visit his garden while in Bishopville, which turned out to be a bonus since I had no idea this was even there. Each year thousands of tourists travel to the garden to marvel at the artistic flora, so I felt fortunate to have stumbled upon it. Not only was the area uniquely serene, but it was also amazing because Fryar literally created this masterpiece in his own backyard—now an official state tourist attraction. Apparently, it all started because he wanted to win "yard of the month" in Bishopville! It also gave me reason to laugh. A finely manicured topiary garden couldn't have been any more polar opposite from tales of a monster running amok in a swamp.

But just a few miles away, Scape Ore Swamp sits on the edge of Bishopville where the infamous Browntown Road extends west to the rural community. It is here that the small town of Bishopville ends and the shadows begin. Even today, Browntown is a mere spattering of houses dispersed along the swamp basin and vast woodlands that spread out into the greater part of Lee County. It's not surprising that no one ever claimed to see the alleged Lizard Man in Bishopville itself. Browntown, however, is another story, as we have learned. Just driving into this area, it becomes apparent that if something were to hide in the South Carolina countryside, this would be a perfect place to do it. At the edge of the swamp, here in Browntown, there are no fanciful spiral shrubs or award-winning lawns. There is only the beginning of a dense fortress of unbridled nature where all things suddenly seem possible.

The Name Game

When someone mentions the word "swamp," most people have a standard mental image of the landscape, usually derived from a combination horror movies and nature shows. Dark waters, drooping trees, wispy moss, prehistoric-looking plants, and even fog are all part of this collective image. To a certain extent this is correct, but not all swamps are alike. Depending on how the water flows—or doesn't flow—will ultimately determine what kind of plants or trees will grow in the area, and what color the water will be. For example, Shrub Bogs are dominated by dense shrubbery, but may also have pond cypress or bay trees growing in the water. Depression Swamps are low areas flooded by rain, runoff, and only occasionally by streams. Various types of oak are usually present there. Basin Swamps are large, sandy depressions near the coast that have filled with water. In these areas, pond cypress, swamp tupelo, and swampbay trees stand among vast floating islands of peat moss.

Aerial view of Browntown Road near Scape Ore Swamp (photo by Liston Truesdale)

Another category is River Swamps. These bottomland areas are found along rivers and creeks, or alongside sluggish streams. When the swamp is fed by a sluggish stream, it is specifically categorized as a Blackwater Swamp. Bald cypress trees, the kind that are often associated with our standard image of a swamp, are common in these areas, along with dark, brackish water. Somehow it is not surprising that Scape Ore is classified as a Blackwater Swamp.

With all of its waterways, Scape Ore Swamp occupies nearly 179,000 acres of the Sandhills and Upper Coastal Plain regions of South Carolina. The main Scape Ore tributary originates in the northwest corner of Lee County, near Lucknow, and flows southward through Browntown, eventually crossing into the neighboring county of Sumter. The Scape Ore channel continues to flow southeast through Sumter until it joins the Black River some 16 miles south of Bishopville. The combined waters then drain in a southeasterly direction until they merge with the Pee Dee River and empty into Winyah Bay at Georgetown, South Carolina.

As the Scape Ore basin meanders through the countryside, thick canopies of southern pine, cottonwood, maple, black gum, and oak rise up from its hillslopes and upland areas. These riparian woodlands, which make up nearly 60 percent of the area, are interspersed with crop and pasture lands sprawled along on the broad summits of the Sandhills. Down in the flood plains, lowland hardwoods crowd the waters with an assortment of bald cypress, swamp tupelo, water tupelo, and water oak to give it that classic swamp look.

The basin land is almost exclusively owned by private citizens; only about four percent of it is owned by Federal, State, or County governments. As such, access is mostly limited to private entries or via the tributary where fishing is common. Plenty of roads run through the area, but because of the swampy conditions along most of the stream channels, routes tend to follow the summits and do not parallel the swampy drainages. With its rich and diverse topography, the Scape Ore Swamp basin can be striking beautiful by day, but after nightfall, it can be a very dark and ominous place.

The origins of the name "Scape Ore" seem to be shrouded in just as much darkness. Over the years the swamp's name has been

Within Scape Ore Swamp
(photo by the author)

noted in land deeds and on maps as *Scape O're, Scape Oar, Scape Hoar, Scapehoar,* or, yes, even *Scape Whore.* The most popular theory of how the swamp got its name directly attributes it to the tale of an "escaped whore," as told to historian Thomas Stubbs by the late Frank McLeod, a local Judicial Court solicitor. As the legend goes, around 1761 a woman of ill repute was causing a stir in the St. Mark's Parish Province of Craven County. Unwilling to tolerate such frisky business in their district, the people of the parish decided to run the whore out of town, or possibly deal her an even worse fate. The woman took heed and managed to escape into the nearby swamp basin, never to be seen again. The spot was promptly tagged with the moniker of "Escaped Whore Swamp," a name that apparently stuck, although it was eventually modified and shortened to "Scape Ore" by officials who felt that the name was a bit too racy.

Another legend attributes the name to an incident that occurred during the American Revolution. According to contributor Robert Cooper in the 1965 book *Names in South Carolina*, published by the University of South Carolina:

> Scape 'Ore Swamp, located near Bishopville, was originally named Escaped Whore Swamp by a group of Revolutionary Volunteers. These soldiers, part of Marion's Brigade, surprised an encampment of British Regulars who were in the process of entertaining ladies of rather shady backgrounds. The British were captured and the Volunteers allowed the terrified women to flee into the Swamp.

The story of the Volunteer heroes is a bit more flattering than parishioners chasing a woman into the swamp, but the debate over which version is correct appears to be moot when looking at previous land ownership documents. The deed for a 100-acre land grant on record at the South Carolina Secretary of State's office, dated July 15, 1768, specifies the area as "Scape Hoar Swamp." This is obviously a different spelling than "Escaped Whore" but still confirms that the name had been established prior to the American Revolution, which didn't start until 1775!

A later deed, found in Book A of the Sumter County Conveyances, notes the transfer of this same 100 acres. In this document, dated March of 1803, the swamp's name is written as "Scape Hore." This is yet a third version (Escaped Whore > Scape Hoar > Scape Hore), but still maintains the idea that an *escape* and a *whore* possibly influenced the original naming. In later documents and on maps, the name continued to be modified slightly, appearing as "Scape O're" and "Scape Oar," until it finally arrived at its modern spelling of "Scape Ore." The modification could have been on purpose to hide the shameful origins, as some local scholars suggest, or accidentally altered due to phonetic spelling errors. The name "Scape Ore" is not inherently easy to spell if one simply hears the name spoken, so it is possible that the spelling changes were due to mistakes. In fact, while talking about the subject of this book with someone who had read my book on the Fouke Monster, they initially thought I said "Skateboard Swamp." I politely corrected them, adding that it would be extremely difficult to skateboard in Scape Ore, even if you were a Lizard Man.

Another interesting theory as to how the swamp got its name is, ironically, related to an actual lizard. According to some local historians, the name "Scape Whore/Hore/Hoar" was a phonetic interpretation of *Sceloporus*, the scientific genus name for a variety of spiny lizards commonly found in the United States. Scape Ore in particular is home to a breed known as the alligator lizard or eastern fence lizard (*Sceloporus undulates*), so it certainly seems like a viable theory. However, scientific classification of the Sceloporus lizard didn't take place until 1802, nearly 34 years after the name "Scape Hoar Swamp" first turned up on a land deed. So while it's truly fascinating to think that the swamp might have been named for its reptilian denizens, it doesn't appear to be the case.

Sceloporus

Regardless of how the swamp eventually got its name, the natives of the area recognized the lizard's significance to the land long before the current sightings. South Carolina was full of Indian tribes such as the Cherokee, Congaree, Cusabo, Natchez, Pee Dee, Santee, and Wateree, all of whom respected these reptiles and often used them for the benefit of their peoples. The Cherokee in particular were known to have used the eastern fence lizard as part of their medicinal treatments and protection rites. In his book *The Sacred Formulas of the Cherokees* (1891), author James Mooney writes:

> There are perhaps half a dozen varieties of lizard, each with a different name. The gray road lizard, or *diyâ'hälĭ* (alligator lizard, *Sceloporus undulatus*), is the most common. On account of its habit of alternately puffing out and drawing in its throat as though sucking, when basking in the sun, it is invoked in the formulas for drawing out the poison from snake bites. If one catches the first *diyâ'hälĭ* seen in the spring, and, holding it between his fingers, scratches his legs downward with its claws, he will see no dangerous snakes all summer. Also, if one be caught alive at any time and rubbed over the head and throat of an infant, scratching the skin very slightly at the same time with the claws, the child will never be fretful, but will sleep quietly without complaining, even when sick or exposed to the rain. This is a somewhat risky experiment, however, as the child is liable thereafter to go to sleep wherever it may be laid down for a moment, so that the mother is in constant danger of losing it. According to some authorities this sleep lizard is not the *diyâ'hälĭ*, but a larger variety akin to the next described.

The next lizard described by the Cherokee in Mooney's manuscript, while not an upright humanoid lizard, does evoke a slight chill when considering what people would report in the countryside of South Carolina more than a century later.

> The *gigä-tsuha'`lĭ* ("bloody mouth," Pleistodon?) is described as a very large lizard, nearly as large as a water dog, with the throat and corners of the mouth red, as

> though from drinking blood. It is believed to be not a true lizard but a transformed *ugûñste'lĭ* fish [a fish said to have horns] on account of the similarity of coloring and the fact that the fish disappears about the time the *gigä-tsuha'`lĭ* begins to come out. It is ferocious and a hard biter, and pursues other lizards. In dry weather it cries or makes a noise like a cicada, raising itself up as it cries. It has a habit of approaching near to where some person is sitting or standing, then halting and looking fixedly at him, and constantly puffing out its throat until its head assumes a bright red color. It is thought then to be sucking the blood of its victim, and is dreaded and shunned accordingly.

The reptile described here is obviously a large, yet normal four-legged lizard, but both the Cherokee and the Alabama tribes also had tales of lizard-like *men*! These beliefs were first recorded by the Spanish explorer and slave trader Lucas Vázquez de Ayllón, who ventured into the Carolinas in 1521. He and a partner landed at the "River of St. John the Baptist"—which is believed to be the Pee Dee River—where they kidnapped 70 natives. One of these natives, given the name Francisco de Chicora, provided some ethnological information about his province, Chicora, and the neighboring provinces. (Chicora was evidently one of several Carolina Siouan territories.) Chicora's information, by way of Ayllón, was included in John R. Swanton's book, *Early History of the Creek Indians and Their Neighbors* (1922). He writes:

> There is another country called Inzignanin, whose inhabitants declare that, according to the tradition of their ancestors, there once arrived amongst them men with tails a meter long and as thick as a man's arm. This tail was not movable like those of the quadrupeds, but formed one mass as we see is the case with fish and crocodiles, and was as hard as a bone. When these men wished to sit down, they had consequently to have a seat with an open bottom; and if there was none, they had to dig a hole more than a cubit deep to hold their tails and allow them to rest. Their fingers were as long as they

Sharp-tail of Inzignanin

> were broad, and their skin was rough, almost scaly. They ate nothing but raw fish, and when the fish gave out they all perished, leaving no descendants.

These strange men, who were also called "sharp-tails" in various native mythos, were said to have possibly come from the sea. They were never directly compared to lizards, but their characteristic tail and scaly skin certainly bring to mind such a comparison. Given the limited writings, it's hard to determine whether these "men" truly existed outside the realms of myth, or even where Inzignanin was actually located, but it's certainly remarkable that natives were reporting scaly humanoids in the North/South Carolina area nearly five hundred years ago.

MORE MYSTERIES

The mysteries of Scape Ore Swamp are not exclusive to the Lizard Man. Over the years the swamp has been a rumored site of ghostly happenings and other spooky weirdness. One of the more well-known examples is the case of the haunted ax. The ax, which had been used to murder a Mrs. M. M. Brown in 1936, served as the woman's grave marker at the Browntown Primitive Baptist Church Cemetery. According to Truesdale, Mrs. Brown was an older woman who lived in a modest home in Browntown near Scape Ore Swamp. In the fall of 1936, she hired a local handyman to perform some chores around the house, including splitting logs for firewood. At some point the handyman decided he didn't want to work so hard for the money, so he grabbed the ax, walked into the house, and proceeded to chop Mrs. Brown into bits. He then stuffed her dismembered body into the fireplace and tried to incinerate the remains. That didn't completely destroy the bones (nor did it get rid of the blood splatter on the floor), so he decided to go ahead and burn down the entire house. He calmly took a total of $36 from the woman's purse, set the house on fire, and walked away from the scene.

A short time later, neighbors reported seeing smoke through the woods, which brought police and firefighters to the scene. The house was not completely burned, however, so the officers were able to reconstruct what had happened. It wasn't long before the handyman's name came up, and they found him working again near the swamp. (Apparently, he was not able to retire on $36.) When they approached him, the man was rumored to have said "I don't know anything about that old lady."

The man was promptly convicted of the crime and later executed in the electric chair. As a gruesome memento, the bloody ax, which had survived the fire because it was completely made of iron, was placed on Mrs. Brown's grave. In time, urban legends developed, claiming that the ax would rotate above the grave on Halloween night, or that it would bring sickness to whomever touched it. It remained at the grave for decades, but just prior to the Lizard Man incidents, the ax was stolen. The first time, the police managed to track it down and return it to its rightful place at the grave. The second time, Truesdale informed the media about the missing ax and a story was run in the newspaper. A short time later, an unidentified woman who had read the story called police to report that she found a bulky, all-iron ax in a trash can on Law Street. Truesdale confirmed that it was the ax in question and playfully theorized that perhaps someone did get sick and that's why they had dumped it in the trash.

Since the ax was being stolen with increasing frequency, Truesdale contacted Mrs. Brown's grandson in Sumter and told him that unless he could mount the ax permanently to prevent theft, they might as well not place it back on the grave. The grandson agreed and Truesdale kept the ax in police custody for 15 years, until he retired in 1993. Since then, the ax has gone missing and its whereabouts are currently unknown.

Another ghostly tale associated with the swamp is that of the "chained dog." It's said that on certain nights, this non-corporeal canine can be heard running across Scape Ore bridge, dragging its broken chain on the road. In an article published in *The Rock Hill Herald* on July 22, 1988, a retired farmer by the name of Romaine

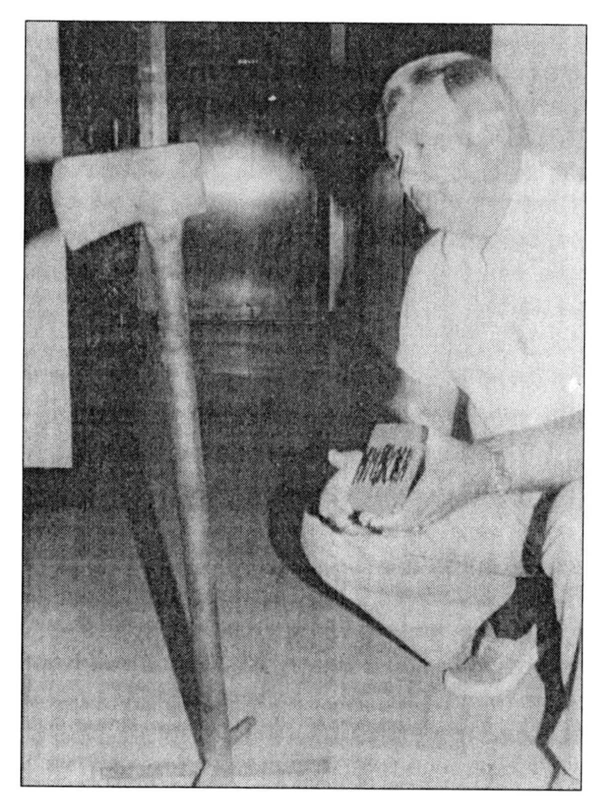

Sheriff Truesdale with the "haunted" ax (photo courtesy of Lee County Observer)

Davis told reporters: "Me and my boys were down there fishing one time, standing on the side of the road by the bridge." At some point during the evening, Davis said he heard a "rattling" on the road. "My son said he saw a dog dragging a chain down the road. I heard the chain, but I couldn't see the dog. My other boy said he couldn't see it either," Davis explained. He didn't know what to make of the incident, but he had always thought the swamp was "spooky."

An even stranger legend is that of the Bishopville Giant. This being was said to have stalked the countryside in May of 1871. According to historian Ann Gregoire in her book *History of Sumter County, South Carolina* (1954):

> The Sumter News described a monster in human form, about 30 feet tall, which walked through the night in Bishopville, plucking fence rails from the roadside. Around the neck was an iron chain from which hung a blacksmith's anvil and sledge hammers, jangling as it moved, breathing like the wind in the tree tops and, at intervals, emitting sounds like the loud pouring of liquid from a bottle.

Following some brief sightings in Bishopville, the strange giant was apparently seen in the town of Manning, some 40 miles away. Records of this sighting can be found in the May 25, 1871, edition of *The Daily Phoenix*. The account begins:

> On last Friday evening, between 9 and 10 o'clock, the citizens of Manning, the County seat of Clarendou, were startled by the sudden appearance in the street of a gigantic being or apparition, which, from the description of an eye-witness, we judge to have been the same that appeared at Bishopville. The thing, or Grand Cyclops, whatever it be, was seen by many persons, both white and black.

According to one witness, as the giant stalked through town, it was able to vanish into thin air and reappear several streets away. It was said to have eventually made its way to the Manning courthouse square where it "disappeared into the well." However, another eyewitness claimed that "he met the giant going off in the opposite direction from where it had been last seen, and that it put out its hand and touched him, and that 'the hand was as cold as ice.'"

There is little else to be found regarding the origin or further appearance of this mysterious monster, so it's hard to tell how much is true, exaggerated, or fabricated. It simply remains a strange footnote in the case of the Lizard Man, suggesting that Bishopville and its infamous swamp were no strangers to monster tales dating back more than a century.

5. Return of the Lizard Man

"That's where the creature was supposed to have run across the road," Sheriff Truesdale explained, pointing to a spot only a short distance from our parked car. He had driven us to a location on Browntown Road near the intersection of Hickory Hill where the Blythers, a family of six, had seen a strange animal while passing through the area back in 1990.

We got out and walked across the hot asphalt for a better view. There was an open field nearby, but also a wall of thick, heavy woods fifty yards off. Something could have easily come out of those woods, run across the road, and disappeared back into the trees without so much as anyone seeing it. Unless, of course, they happened to be driving down the road at just the right time.

I was particularly interested in visualizing this sighting because earlier Cindy and I had been able to read the actual police statements provided by three of the witnesses in the case. It's rare that officials take cryptozoological reports seriously enough to follow strict investigation protocol. But that's what Truesdale had done when the Blythers family came to the station to report the incident. He had each one write down everything they had seen, just as detectives might do in a more serious case.

Now, years later, Cindy and I were able to study the vivid descriptions and emotional reactions of the witnesses. These were not memories that had been distorted or exaggerated by the passage of time; these were recollections recorded just one day after the sighting had taken place. We were fortunate to have access to such documents, but in the case of the Lizard Man, it's not really surprising. During his thorough investigations, Truesdale had followed this procedure for most of the sightings. His collection of records still contains, in many cases, both the handwritten witness statements along with a final typed report. Of course, police documents do not validate witnesses as any more credible than those who do not report to the police, but perhaps they do suggest that if these individuals are willing to go on legal record, they are more likely to be telling the truth.

As we returned to our car, I contemplated the Blythers' case. It was one that would shine a new light of clarity on the Lizard Man mystery.

The Blythers Case

More than two years had passed since the Army Corps of Engineers Colonel reported seeing a unknown creature run across a road south of Bishopville in August of 1988. Since then, not a single person had gone on record to report even the briefest sighting. The news coverage had also long since fizzled out and things had returned to normal around the Lee County Sheriff's Office. The trail of the Lizard Man had seemingly gone cold.

This was a definite relief for Truesdale, who had weathered the brunt of the media storm, but it was by no means closure. Even though he was pretty sure there was no reptile man to be found, questions still remained. Was it a hoax? Did a bad case of swamp fever have everybody seeing monsters? He was beginning to think that's all there would ever be to the story. That is, until a family of six saw something near Scape Ore Swamp that they would never forget.

It was July 30, 1990. Bertha Blythers and her five children had just left an eating establishment in Bishopville and were headed back to their home in Camden. Earlier that evening, she had picked up her son Johnny from her mother's house in Browntown before making her way to a burger place on Highway 15. She had all of her kids for the evening, including Johnny (18), Tamacia (11), Christa (5), and two twins boys (4). Bertha was a hard-working single mother with little to spare, but she enjoyed treating her family to a restaurant whenever she could afford it.

As they traveled back toward Camden, they took the familiar

route along Browntown Road heading west. They were in the best of moods, eating French fries and singing to the radio. They were also talking about the Lizard Man. Not everyone had forgotten the incidents, especially those with family living in Browntown.

At approximately 10:30 p.m., they crossed over the interstate and were nearing the intersection of Hickory Hill Road when suddenly a large figure appeared out of nowhere and lunged toward the passenger side of the car. Bertha's oldest daughter, Tamacia, was sitting in the passenger seat with the window rolled down. When she saw the creature, she screamed. Fearing for her daughter's safety, Bertha cranked the steering wheel and swerved the car. She later told police: "I was looking straight ahead going about 25 mph, and I saw this big brown thing, it jumped up at the window. I quickly sped up and went on the other side of the road to keep him from dragging my 11-year-old girl out of the car."

Once they passed the animal, Bertha slowed down long enough for her son to look back.

"Mother," Johnny said in a panicked voice. "It's still there!"

The bulky figure could be seen loping across the road. It walked on two legs, like a human, but was slightly hunched over and much, much larger. Johnny could only see it for a few seconds before it disappeared into the darkness. Bertha did not have time to take a second look herself. She was too busy trying to calm Tamacia, who was sobbing uncontrollably.

In a statement given to the Lee County Sheriff's Office, Bertha described the creature as being tall, wide, and having "two arms like a human." She could only see it from the waist up, but there was no question that it was "big." She was not able to make out any clear facial details in the short time it was in front of her, but she was quite sure the body was covered in brown hair. "I never seen anything like it before," she told the police. "It wasn't a deer or a bear. It was definitely not a person either."

Johnny agreed. In his own statement to police, he described the creature as being "6 feet tall or more," covered in brown hair, and having two big eyes. "I certainly don't think it was a human dressed in something because a human ain't going to jump in front of a

Location of the Blythers' sighting on Browntown Road (photo by the author)

car," he said.

Once the figure disappeared from view, Johnny pleaded with his mother to turn around so they could get another look. But Bertha was not going to do any such thing. "There was no way I would have turned that car around and went back," she said in her statement.

Bertha raced home and got the kids into the house. Still shaken, she called her sister and told her of the bizarre incident. Her sister urged her to call the police as soon as possible. Bertha hung up and did just that.

Deputy Ed Corey from the Lee County Sheriff's Office was dispatched to the scene. It had been a while since one of Truesdale's men had responded to a monster report, but it all came back quickly enough. The officer searched the area with a flashlight but was unable to find any trace of the large, hairy creature. The grassy areas on both sides of the road were scanned for tracks, but again he came up empty handed. Whatever had frightened the Blythers had simply vanished without a trace.

The following day, Bertha, Johnny, and Tamacia came to the Sheriff's office where they gave statements to Sheriff Truesdale.

The Blythers family has a close call with a huge, hairy creature

As with any proper investigation, the witnesses were interviewed separately so that their stories could be compared for inconsistencies. After each one had told their side of what happened, Truesdale could find no reason to doubt them. None of them wanted anything to do with television interviews or other media recognition; they only wanted to tell the truth about what they had seen. Johnny, especially, was familiar with the ups and downs Chris Davis had suffered due to his sighting, and he wanted to make sure that did not happen to him or his family.

As in the Davis case, Truesdale could not be sure *what* the witnesses had seen, but he had no doubt they had seen *something*. That left only two possibilities. Either they fell victim to a hoaxer… or the Lizard Man was still out there.

Another Close Call

A short time after the Blythers' sighting, a bank vice president from Florence allegedly told several people that he had a strange encounter while deer hunting near Scape Ore Swamp. Apparently, he had seen a figure that fit the general description of the Lizard Man prowling around his tree stand. It seemed like another promising lead, but he would not come forward with more details due to his prominent position at the bank.

It wasn't until1992 that another report of any substance came to light. In May of that year, Brian and Michelle Elmore, a couple in their early twenties, approached Sheriff Truesdale with a startling story. According to the Elmores, they had nearly collided with a large, gorilla-like animal while driving near Browntown more than a year earlier. Being from the area, they were naturally acquainted with the Lizard Man stories and felt that the creature might be the same as what others had reported. The incident had truly frightened the young couple, but they decided not to go public with the story at that time for fear of ridicule. Over time, however, they came to the decision that it might be a good idea to tell Sheriff Truesdale of their encounter.

The incident had taken place in the fall of 1991. The Elmores were headed north on Gum Springs, a road that connects the communities of Browntown and Cedar Creek. It was around 12:30 a.m. and they were in route to a late-night party. As they proceeded through the darkness, Brian's attention was suddenly drawn to the side of the road where he saw some sort of animal crouching in the headlights. According to Brian's signed statement:

> All of a sudden, I saw something standing straight up on the side of the road. I asked Michelle, "Did you see what I saw?" It had its arms over its head. She said it must have been a cow. I said I didn't think it was a cow. So we turned around and went back toward the place where I saw it. When we got near it, the lights hit it. Michelle yelled, "Look don't hit it!" It was running in front of the truck. I swerved to miss it or I would have hit it. It ran in front of us and then ran into the woods.

Brian described the creature as looking "like a gorilla... but a lot bigger." He was also sure that it didn't have a tail. "It had arms and legs and ran humped over," Brian wrote in the report. "Never seen anything like it before or since. I've been to the zoo and the circus, I've seen gorillas and bears—this was definitely not a gorilla or bear."

Michelle's own report echoed the same startling details as her husband's:

> We were riding along and saw something on the side of the road humped over like it was picking up something, or balled up like it was hiding, I guess when it saw the light. All of a sudden it jumped up and ran straight in front of us and into the woods on the opposite side. If Brian hadn't hit the brakes and swerved over, we would have hit it.

She described it as being seven feet tall or more, brownish-black in color, with ears about "mid-way down" on its head. She also noted that it had very long arms and "ran humped over like a gorilla would run." She did not describe it in terms of a reptile, but in terms of another, more familiar cryptid:

> I know what I saw, the car lights were on it and it was very plain. At first we thought it was a cow. I didn't see it but Brian said it was standing straight up with its arms over its head, but we turned around... that's when I saw it and realized it was a Sasquatch.

The sighting completely turned their night upside down. Seconds after the thing disappeared into the woods, Brian and Michelle dropped the idea of attending the party and instead raced to Brian's parents' house where they told them of the incident. In time they told a few others, but as often happens in these types of incidents, they were met with a mixture of doubt and laughter. "The reason I didn't report it is because everybody would think I was crazy, and everybody kept saying it was a bear, so I left it at that," Michelle explained. "It looked exactly like a Sasquatch to me."

The day after the sighting, Brian returned to the scene, bringing along a trusted friend. The two men proceeded to look for any hard evidence that might corroborate what he and Michelle had seen, but unfortunately, they came up empty handed. The grass and hard clay along the side of the road was simply not conducive to footprint impressions, and they could find no hair or other signs of the creature's passage.

Like Michelle, Brian resolved to forget about the incident and move on. What else could they really do? It wasn't until they befriended Truesdale that they decided to reveal what they had experienced that night. A year and a half had gone by, but they were still adamant about what they had seen. Their seriousness is readily apparent in their statements.

Truesdale not only found the couple credible but also found details of their sighting to be highly consistent with those provided by the Blythers family. The incidents were so similar, in the description of the animal itself and in the way it moved across the road, that it led one to wonder if the families knew each other or the Elmores had somehow read the Blythers' earlier witness statements.

But that was not possible. The witness statements of the Blythers had never been made public, nor did the Elmores know the Blythers. It was either pure coincidence, or corroboration that a sasquatch-like creature was haunting the areas around Scape Ore Swamp.

Vehicular Carnage

In the years following the Blythers and Elmore reports, other strange stories often got thrown into mix as newspaper journalists would occasionally churn out retrospectives on the Lizard Man Summer of 1988. In one case, it was reported that in 2004 some kind of scaly creature tried to snatch and drag a young girl into a river as she walked by. The incident was attributed to the Lizard Man but sounds more like a close encounter with a reptile of the real kind: an American alligator. This juicy tidbit is often mentioned on websites and in blogs, but in no case is it ever backed up with any references or other proof that this incident took place anywhere near Scape Ore Swamp, or in fact, ever took place at all. If it did, there are certainly numerous alligators living in and around Scape Ore Swamp and its associated tributaries, so it would not be surprising if the culprit in this case was simply a hungry gator.

Another incident which is often lumped into the Lizard Man story involves a dubious sighting in Newberry County, approximately 80 miles from Scape Ore. According to *The Newberry Observer* in 2005, a woman reported seeing not one, but *two* strange creatures outside of her home one evening. She called police, but the officer dispatched to the scene could not find any evidence to corroborate the story and ultimately decided that her sanity was questionable. To comfort her, he told the woman "they just like to check on humans from time to time."

Aside from these thinly associated stories, the Lizard Man case lay dormant until 2008 when a new incident literally brought the whole thing full circle. It all started when Bishopville resident Bob Rawson walked out of his front door on the morning of February 28. He was headed to the local hardware store to pick up something for his wife. As he approached their mini van, however, he realized their day might not go as planned.

Their sparkling blue Dodge Grand Caravan was in a bizarre state of damage. The front edge of the hood had chunks missing, as if something had chewed on the metal; the grill had a series of puncture holes that resembled teeth marks; the edges of both front fenders were heavily scratched and bent; and to top it off, there was blood on the front and side of the van. It literally looked as if the car had been "attacked" by a huge, angry animal.

Bob summoned his wife, Dixie, who also looked at the car. She was equally confused by the nature of the damage. As they began to inspect the rest of the area, they found that their previous day's newspaper had been shredded and strewn in the street, along with some towels taken from boxes on the side porch of their home. With all that activity, plus the automobile mauling, why hadn't they heard anything during the night? None of it made any sense.

Confused and alarmed, the couple placed a call to the Lee County Sheriff's Office. Officers were sent to the home where they began an investigation into the matter. No one wanted to be the first to say it, but there was no denying that the incident was eerily similar to the one that initially set off the Lizard Man craze 20 years earlier. The police had never been able to reconcile the bizarre damage to the Waye's vehicle, and they certainly couldn't explain the situation with the Rawsons' van, especially since the damage was far more extensive. If it were an animal, what kind of animal could bite through steel as if it were tin foil?

Another thing the Rawsons' noted was the disappearance of several cats, which normally slept on their side porch. Mrs. Rawson estimated that there were up to 20 cats that made their home there, but on the morning the van was found damaged, at least half of the felines were missing. One theory was that a predator was after the cats. When the cats fled under the van for safety, the predator became enraged and tore up the vehicle. "I had never seen anything like this," Bob Rawson stated in the March 4, 2008 edition of *The Item*. "It looks like something got after the cats and they went under the van."

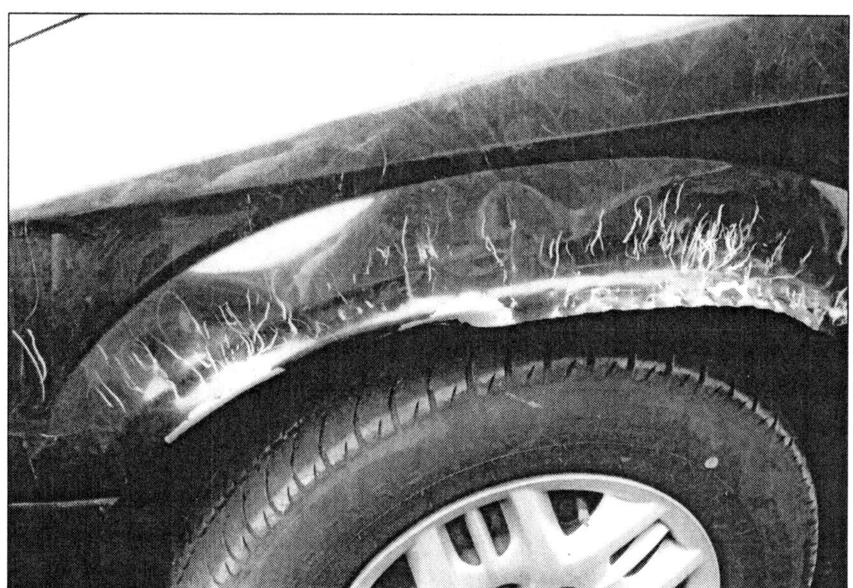

The Rawson's damaged van
(photos by Liston Truesdale)

Dixie Rawson entertained the theory of a predator, but still had her doubts. "What I don't understand is whatever it was got the towels out of the boxes where the cats sleep and didn't disturb the boxes," she told reporters.

The Rawsons' son-in-law laid the blame on a bear. Others hailed the return of the Lizard Man (although that didn't include the Rawsons, since apparently they had never heard of the Lizard Man). The police weren't above considering their old nemesis, but their best course of action was to take samples of the blood and send it off for DNA analysis. Since the year of the Waye's car damage, science had come along way with DNA technology, providing the police with far better tools. According to Major Danny Simon in *The Item*: "We are sending the blood samples to the State Law Enforcement Division."

While the sheriff's office was waiting on the test results, Sheriff E.J. Melvin decided to take a four-wheeler into the fields near the Rawsons' home to look for further evidence. It wasn't long before he discovered the bodies of two dead animals: a cow and a coyote. Their corpses were not far from the Rawson's van.

Given this, Melvin theorized that the coyote had been responsible for the vehicle attack and the disappearance of the cats (which were still unaccounted for). In the same article, however, Mrs. Rawson brought up a good point: "If the coyote did this damage, what killed the coyote and cow?" The dead animals only added to the mystery.

By May 15, the DNA results were in. According to a report submitted to the Lee County Sheriff's Office by Elizabeth Wictum, Forensic Laboratory director at the University of California School of Veterinary Medicine in Davis, "the likely culprit" was a domestic dog.

A dog?

The results were definitely anti-climactic for those believing that the Lizard Man might be reponsible, but even those who had their money on something more mundane, like a bear or wolf, were at odds with the findings. In an article published in *The Item* on May 15, 2008, Sheriff Melvin stated that he could not accept the report's findings. "If I didn't see the van, I might believe it," he told reporters. "But I still don't think a dog is that strong. My theory is that it belongs to the canine family, but it is a coyote or wolf."

The Rawsons also disagreed with the findings. "I just don't believe a dog could do that," stated Bob Rawson. Dixie concurred: "I can see a bear doing this but not a dog."

To account for a misidentification, the Rawsons offered a possible explanation. They said that: "Before the swabs were taken from the van, [they] visited a neighbor who has three German Shepards." They theorized that the shepards somehow contaminated the samples when they sniffed around the blood stains.

Others, such as Janson Cox, director of the Cotton Museum, were not so quick to dismiss the dog explanation. Cox had overseen the Animal Forest Wildlife Zoo for 25 years while he was curator of Charles Town Landing, a popular South Carolina state park. Cox told reporters from *The Item* that "a wild animal will typically not do anything to endanger its life such as tearing something with its mouth and teeth. But a domesticated animal will." In his opinion, the culprit could have been a pit bull.

No matter what the DNA suggested, or what animals were known to be in the area, the core question seemed to be: could any known animal actually do that kind of damage to a car? If not, then it only left two possibilities: (1) that it was done by humans—hoax, vandalism, or otherwise—or (2) that it was done by some yet-identified creature lurking in the woods near Bishopville. It seemed that a more thorough scientific test was in order, but the people of Bishopville would have to wait several more years before any such test would be done.

One More Bite

While most Americans were firing up the grill and getting ready for an evening of hot dogs and fireworks, Leon Marshall was trying to figure out what the hell happened to his car. On the morning of July 4, 2011, the Bishopville resident walked to his driveway only to find that his 2009 Dodge Journey had been extensively damaged during the night. According to a report from the *Lee County Observer*, they found "teeth prints" on both fenders that went completely through the metal, and a loose bumper with its underlying material torn out and scattered on the ground. "It was a shock," Leon's wife, Ada, told reporters. "My husband was yelling for me to call the police because something had torn the car up."

When Ada placed the call, the Lee County Sheriff's Office responded, knowing full-well what they would be in for. The responding officer, Deputy Fletcher Williams, had been the first officer on the scene three years prior when the Rawsons reported their own mysterious car damage. Now it was like a bad case of déjà vu.

At the scene, Williams inspected the car and surmised that teeth were responsible for the damage, as if the car had been "chewed up." To further suggest that an animal had been the vandal, a clear trail of saliva was visible from "the front passenger door across the front of the vehicle to the driver door" and on the discarded strips of bumper material. He also found long white and brown hairs.

Once again, the evidence seemed to point towards some kind of a canine attack. No one could figure out why dogs would want to do such a thing, or how they could bite through metal, but it was the best explanation the police could offer without further testing. All they could say for sure was that whatever was mauling the cars in Bishopville seemed to have an uncanny taste for Dodge and a knack for getting away undetected.

The Lizard Man was, of course, brought up again. But without a corresponding sighting, or even any recent sightings, this was merely part of the Lee County judicial process when trying to figure out strange occurrences. Deputy Williams reportedly asked the Marshalls straight out if they thought the Lizard Man was to blame (mostly likely in jest). After all, they did live in close proximity to the swamp. "Our neighbors are afraid to go out after dark since they don't know what is responsible for the damage and Scape Ore Swamp is right across the road," Ada told the *Lee County Observer*. "I've lived here all my life and never had anything else like this to happen."

The police had not benefited much from the DNA test in the previous case, so they decided to forego the expense and simply chalk it up to animal vandalism. It could have been an intentional act by the Marshalls for insurance fraud or Lizard Man glory, but this didn't seem likely. The Marshalls were just not the type of folks who would suddenly start pulling scams, nor did they seem to be in it for attention. In fact, Cindy and I tried to interview the Marshalls during our visit to Bishopville, but they denied our request. They are simply tired of talking about it and wish to move on.

The Marshalls, however, did grant a previous on-camera interview as part of the SyFy show *Fact or Faked: Paranormal Files.* In the Season 2 episode entitled "Reptile Rampage," the team of paranormal investigators visited Bishopville in an attempt to determine if the Lizard Man sightings and/or car damage reports were real or "fake." As part of the investigation, the team talked to Leon and Ada Marshall, and studied photos of their damaged Dodge vehicle. The team then put three theories to the test in an effort to determine if the damage had been made by an animal, bullets, or by tools. The police had, of course, performed DNA tests in the Rawson case, but these new tests were the first to analyze all the possibilities, including a hoax or vandalism.

To find out if an animal could have done the extensive damage, the team first spoke to a wildlife specialist who agreed that an animal, such as an alligator, may have the bite-power required to chew up a car. The team then fashioned a hydraulic jaw that could be used to replicate the bite marks as shown in the vehicle photos. This test fell short, however, when it did not produce results like those seen in the photos.

The next test centered around whether bullets were the cause. The team proceeded to fire two types of guns at the hood of a car, but the results were not a match either. The bullet holes were similar to the "bite marks" in the photo, but appeared too clean. The marks on the Marshall's car were much more "scratchy" as if something had "chewed" the metal, not simply pierced it.

In the third test, the team evaluated whether power tools might have been used to fake a "creature attack." The three team members gathered an assortment of drills and sanders, and were then given one minute to see how much damage they could inflict on the car. The results in this experiment were very close to what appeared in the photos, suggesting that perhaps human vandals were to blame.

As a final part of the investigation, the team performed a "sociological hoax experiment." To do this, they dressed up one of the team members as a "red-eyed reptilian monster" and shot a video of "it" creeping through the woods. They posted the video online and then monitored the comments. Some of the comments indicated skepticism, but "the majority suggested that there is a deep-seated belief in the Lizard Man." This led the team to conclude that all of the eyewitness sightings were a misidentification or a hoax.

These tests are undoubtedly worthwhile and were performed in a reasonably scientific fashion, but in the end they had some shortcomings. First, the Hydraulic Jaw test could only *replicate* the actions of an animal, not truly show what a live animal could do to a vehicle if enraged or otherwise provoked. Given the presence of hair, saliva, and tracks at the crime scenes, it strongly suggests that an animal was responsible for at least one, if not more of these incidents. Just because a mechanical stand-in could not recreate the damage, does not mean that an animal could not. Animals have been known to do amazing things.

The Power Tool test was the only thing that came close to recreating the exact damage found on the Marshalls' car, but the only way this could have occurred was if the owners had hoaxed the attacks. If it were done by a marauding band of car molesters who pulled up in the middle of the night and assaulted the car with loud power tools, the car owners or their neighbors would have surely heard or seen something. In no mystery car mauling case—Waye, Rawson, or Marshall—did anyone hear anything during the night. This pretty much means that the only scenario in which this could happen is if the owners inflicted the damage on their own cars, which makes very little sense. In the end, it seems that the animal theory still holds up best, whether it was an ordinary animal or one that has yet to be discovered.

As far as the Sociological Hoax Experiment being used to dismiss the entire history of Lizard Man sightings, this conclusion seems much too broad. While the effects of rampant "Lizardmania" may have influenced some of the Lizard Man sightings, the conclusion does not take into account the fact that several of the Lizard Man witnesses claimed to have seen the creature *before* there was public knowledge of it. And, as we have learned, the majority did not even describe the beast in reptilian terms. It would seem, rather, that if the original witnesses were predisposed to believe in a walking lizard man, then their descriptions would have more closely matched the public's preconceived ideas of such a creature. Yet this is not the case.

Another issue concerning this broad conclusion is that the test was conducted over the internet. In this case, there is no way to attest to the credibility or sanity of the people who believed it to be a real video of a "lizard man." These people, unlike the actual eyewitnesses, could not be interviewed by the press, the police, or myself. They were simply anonymous persons making comments based on a fancy television hoax.

In the end, there is only one thing this hoax experiment can tell us with any certainty. And that is: people are still fascinated by the Bishopville Lizard Man.

6. Other Monsters From the Mire

The Bishopville Lizard Man may be the most famous of cryptozoology's purported scaly humanoids, but it's certainly not the only one of its kind. Sightings of similar creatures have been reported elsewhere, although not in great numbers or with consistent descriptions. Some of these accounts describe a man-like reptile, while others describe something of a more amphibious nature. These creatures are often seen near watery environments such as rivers, lakes, and swamps, thus earning them the general classification of "swamp monster" or "creatures from the black lagoon."

To be fair, the term "swamp monster" may also be applied to a range of cryptozoology creatures, not all of them having scales, fins, or gills. For example, some hairy hominoids are said to dwell almost exclusively in bottomland habitats. These creatures, which include cryptids such as the skunk ape (a.k.a. swamp ape), the Honey Island Swamp Monster, and the Fouke Monster, would fall more into the sasquatch/Bigfoot category despite the tendency to assume that all swamp beasts would be of the green, slimy sort.

That being said, the cases of *scaly* humanoids becomes more intriguing. Whereas we have thousands of reported sightings of sasquatch-like creatures around the world, and likewise for various lake serpents, reports of scaly humanoids are found in smaller numbers. Perhaps these type of sightings are so frightening and unbelievable that people tend not to report them, or these creatures are far more elusive and mysterious.

Because of the strange nature of scaly man-like monsters, cryptozoology has struggled to identify and classify just what sort of creatures these may be. Ivan T. Sanderson, one of the first recognized cryptozoologists, included a few sightings of scaly humanoids in his larger tomes of cryptids, but placed them alongside reports of hairy, ape-like creatures. Paranormal researcher and author John Keel did the same. In his book, *Strange Creatures from Time and Space* (first published in 1970), Keel presents reports of swamp creatures in a chapter titled "Creatures from the Black Lagoon." The chapter includes monsters of the scaly sort, but also hairy hominoids that are more akin to Bigfoot than Universal's Gill Man. However, since these cases tended to occur within close proximity to water, he brands the whole lot with the tongue-in-cheek name of "Abominable Swamp Slobs," or A.S.S. for short. (A term derived from Abominable Snowmen, or ABSM, which was used at the time to describe all Bigfoot or Yeti-like creatures.)

Both Sanderson and Keel borrowed heavily from the files of cryptozoologist Loren Coleman. One of the most prominent and widely recognized figures in the field today, Coleman was the first to research swamp monster cases, starting more than four decades ago. He wrote of them in his early articles for *Fate*, *Fortean Times*, and *Strange* magazines, and later in his books and blog entries. In *Fortean Times* articles in the early 1970s and in his books *Mysterious America* (1983) and *Curious Encounters* (1985), Coleman was the first to begin grouping these cases together using an underlying feature of these cases such as the presence of water. Since many of the sightings took place along specific waterways (for example, the Ohio River Valley) he also grouped them by region.

In Coleman's book, *The Field Guide to Bigfoot and Other Mystery Primates* (1999), co-written with Patrick Huyghe, the authors included a few water-dwelling humanoids under the classification of "merbeings." They suggested these creatures could be divided into two distinct categories: a "marine subclass," which has fin-like appendages, and the "freshwater subclass," which is said to have three toes. According to the authors: "The freshwater creatures are often found venturing onto land and are far more aggressive and dangerous, being carnivorous, than their calmer marine cousins." The authors' classification and theories are contingent upon the theory that these freshwater monsters are actually covered in hair, not scales, which justifies their inclusion in a book about mystery primates. To explain the descriptions of scaly skin, they proposed that "the freshwater variety occasionally has patchy hair growths that appear 'like leaves' or 'scaly.'"

In a later book, *Mothman and Other Curious Encounters,* Coleman includes famous cases of reptilian or amphibious humanoids under the heading of "Lizardmen," which is certainly a logical grouping since most of the creatures exhibit at least some man-like lizard traits. This idea, first formulated by Coleman and author/researcher Mark A. Hall years ago, has become a more popular and simplified term used to identify any such creatures seen in and around bodies of water.

Hall's self-published book, *Lizardmen,* which he released in 2005, also utilized this general classification term. Like Coleman and Huyghe, he theorized that perhaps these scaly humanoids were not lizard-like at all but rather primates that had long ago developed an affinity for living in watery environments. These "water apemen," as he calls them, have been misidentified as mermaids and mermen in times past, and more recently identified as "lizardmen," a term that encompasses the apparent traits as well as the sensational aspects of these creatures.

But no matter what we call them, getting an understanding of what they represent is no easy task given the rarity of sightings and lack of evidence. The only thing that's certain is that people from all over North America have occasionally reported harrowing encounters with upright-walking creatures that seem to be covered in scales or have other reptilian/amphibious traits. Are these real-life Creatures from the Black Lagoon or are they simply a type of primate that has been misrepresented by names such as Lizard Man? Or, are they something completely different and altogether otherworldly? Perhaps by delving into some of the most famous sightings of scaly bipedal creatures we can not only come closer to answering this question, but we can also shed some light on the case of the Bishopville Lizard Man. If such creatures can be shown to exist in other parts of the United States, then perhaps they can also be found within the murky swamplands of South Carolina.

Riverside Reptoid

The date was November 8, 1958, and the hour was late when a man by the name of Charles Wetzel drove his 1952 Buick along North Main Street in the California town of Riverside. He was listening to the radio and enjoying the ride as he approached an area where the Santa Ana River often flooded the road. There, as usual, a shallow plane of water engulfed the lowest portion of the pavement, so Wetzel slowed down.

About the same time, his radio began to pick up static. Wetzel tried another station, but the airwaves only offered white noise. He was about to shut it off when his attention was drawn to a temporary danger sign that marked the flooded area. However, before he could even read the sign, something else far more unexpected commanded his attention. It was a six-foot-tall creature, which had run into the middle of the road. Wetzel hit the brakes. The thing was just standing there looking at him.

It had a "round scarecrowish head," Wetzel told reporters from the *Los Angeles Examiner*, which first ran the story on November 9, 1958. He described the face as having a beak-like mouth and fluorescent, shining eyes with no visible ears or nose. And in both the original interview and an interview conducted in 1982 by Loren Coleman, Wetzel stressed that the creature had skin that was "scaly, like leaves, but definitely not feathers." A scaly humanoid?

Wetzel was shocked by the sight of the creature and for a few moments he and the creature remained transfixed, eyeing each other in the road. Then, without warning, the creature advanced toward the car and began clawing at the windshield. It crawled on top of the hood as it made a "gurgling sound" mixed with "high-pitched screams."

Utterly frightened, Wetzel fumbled for the .22 pistol he always carried with him. He raised it in warning, but the creature paid no heed as it continued to scratch at the glass. Wetzel then made the decision to stomp the gas pedal and let the beast deal with the

Charles Wetzel encounters a scaly humanoid creature in 1958

consequences. As the car lurched forward, the creature fell in front of its path and was run over. Wetzel simply sped away, leaving the creature lying in the road.

Wetzel went directly to the local police where he told them of the incident. The police, taking him at his word, quickly rounded up some bloodhounds and made their way to the area. They searched extensively upon the road and along the river, but could find no evidence of the alleged creature. The only tangible indications that something had been in contact with Wetzel's car were some marks on the windshield and a streak of missing grease along the bottom of the oil pan that seemed to confirm the man had indeed run over something.

In Wetzel's extensive interview with Coleman, he elaborated on some of the details. Among these was his peculiar description of the creature's legs. According to the recap in Coleman's book, *Mysterious America*, Wetzel told him that the creature's "legs stuck out from the sides of the torso, not from the bottom" and that it also had "incredibly long arms."

If Wetzel's story is true, it does appear to be one of the earliest sightings of a scaly humanoid in modern times. Although some Bigfoot researchers, such as John Green, have included this encounter in their sasquatch books, Wetzel was clear that the creature had skin that was more "scaly" than fur-covered. The weird details about the beaked mouth and lack of a nose are also very peculiar, further separating the creature into its own category.

As puzzling as the creature may have been, Wetzel was not the only one who reported a mysterious encounter near the Santa Ana River. The following night, another motorist driving along the same street reported seeing a large shadowy "something" run across the road. Could it have been the very same beast? We will never know.

Loveland Frog

Another famous case, this time involving multiple sightings of frog-like cryptids, occurred near Loveland, Ohio, in the years 1955 and 1972. The alleged creatures, which have come to be known as "Loveland Frogs" or "the Loveland Frogmen," were first spotted in July of 1955 by Civilian Defense volunteer, Robert Hunnicutt as he was driving on a dark stretch of road near Loveland, a small city northeast of Cincinnati. According to researcher Leonard Stringfield, Hunnicutt was traveling at 4:00 a.m. when his headlights fell upon three or four diminutive humanoid creatures standing under a bridge. Curious, Hunnicutt pulled over to observe the entities. He described them as being bipedal, quasi-reptilian beasts about three feet in height with "lopsided chests, wide, lipless, froglike mouths, and wrinkles rather than hair."

Hunnicutt watched the strange creatures for approximately three minutes until one of them held up a bar-shaped device that appeared to be generating blue sparks. At that point, he hit the gas pedal and sped away, noting that he could smell an odor reminiscent of "fresh-cut alfalfa" as he passed by the area. The witness then drove to the police station where he reported the sighting to Loveland Police Chief John Fritz. Although reports vary, supposedly Fritz went to investigate the bridge and ultimately placed an armed guard there. It is also rumored that the FBI looked into the case.

The strange incident generated sparks of its own within the community, but eventually the buzz faded as no other immediate reports came to light. It would be seventeen years before another sighting of such a creature would occur in Loveland. In this incident, which occurred on the night of March 3, 1972, a police officer by the name of Ray Shockey was driving near the Little Miami River when he saw what he believed to be a dog on the side of the road. The conditions were icy, so he proceeded with caution. As he got closer, the animal suddenly got up and darted in front of the car revealing itself to be a bipedal creature with a frog-like face and leathery skin. The officer slammed on the brakes and stopped the car. For a few moments the creature just stood there eyeing the officer before jumping over the guard rail, running down the embankment, and finally disappearing into the waters below.

Shockey estimated the creature's height to have been three to four feet with an estimated weight of 75 pounds. He said the

Loveland Frog

creature was something like a lizard or a frog, although it definitely moved on two legs.

Shockey knew his story was incredible but nonetheless filed a police report. It was an action he would later come to regret due to the heavy ridicule that followed. The derision was so intense, in fact, he eventually stopped talking about the incident. In a 1999 *Cincinnati Post* article, Shockey's mother was quoted as saying "Ray took so much ridicule over that thing that he stopped giving interviews a long time ago."

Officer Shockey may have suffered the ribbing of his incredulous peers, but he was not the only one to see something strange on that road in 1972. A mere two weeks after Shockey's event, patrolman Mark Mathews drove up on some sort of animal as it sat in the middle of the blacktop. Mathews stopped to get a better look, and when he got out of the patrol car, the thing lurched toward him. Mathews drew his gun and fired three shots, hitting the thing. It "leaped convulsively and fell into the river," Mathews told reporters at the time.

Mathews' description of the creature reportedly matched Shockey's with the exception of a tail, which Mathews believed

The bridge where the Loveland Frog sightings took place (photo by John Covington)

he saw on the animal. However, years later, Mathews claimed that news reporters had misrepresented his story. In the *Cincinnati Post* article of August 14, 1999, Mathews told writer David Wecker that: "I got a real good look at it; and it was obviously a pet lizard." He theorized that the creature was an escaped pet that was seeking warm water. "This was in February, and it was hanging out near the Totes [Isotoner Corporation] plant, where the hot water used to spill out into the river, trying to keep warm," Mathews explained.

It's hard to know whether Mathews saw something legitimately unexplainable or whether he changed his story later to avoid ridicule (just as Shockey had done by refusing to speak further on the subject). One obvious question is why would he open fire if it was just a common lizard? Either way, that was not the end of the reports. According to the *Cincinnati Post* article, a high school student claimed to have seen the Loveland Frog two months after the policemen's sightings. The un-named student said the creature was "a green, 150-pound thing that he estimated to be 3 feet, 11 inches long." [7]

Later in 1972, a farmer claimed to have seen four bizarre creatures while working in his field. The creatures were described as having greenish-gray skin, large circular eyes, and a mouth full of sharp teeth. The nervous farmer allegedly observed the beasts for a short time before they headed toward the river and eventually disappeared in the brush.

The case of the Loveland Frog is a peculiar one, given the retracted statement by Mathews and the near unbelievable prospect of a "frogman." One thing that seems highly coincidental is the date of the first reported sighting: 1955. This was only one year after Universal's *Creature from the Black Lagoon* hit the theaters. Could this have influenced the description of the creature given by the witness? It might pierce the bubble of the entire Loveland Frog case if not for the later sighting by Officer Shockey. If, like Mathews claimed, he had merely seen a misplaced pet lizard, then why would he file a police report? It doesn't seem to make sense.

7 The fact that this student could make such a precise height estimate makes this report highly suspicious and should be regarded as such.

Neither does the fact that this particular "pet lizard" was able to walk on two legs!

The year 1955 also registered another strange case in the category of man-like water monsters. It was not in Loveland, but occurred just up the Ohio River Valley in Evansville, Indiana. As the story goes, Mrs. Darwin Johnson was swimming in the Ohio River with a friend when a clawed hand clutched her leg and tried to pull her under the water. Johnson, in a state of sheer panic, fought desperately against her watery assailant, who seemed intent on drowning her. Fortunately, she was able to free herself and swim to the safety of an inner tube.

Neither woman got a look at the creature, since it remained submerged during the entire ordeal, but they were certain that whatever had grabbed Mrs. Johnson's leg had a human-like hand since a palm-print with a green stain was clearly visible on her calf.

Thetis Lake Monster

Three incidents concerning another strange creature from the depths took place at Thetis Lake in British Columbia. The first, as reported by the *Victoria Daily Times*, occurred on August 19, 1972 (the same year as the major Loveland Frog sightings). Two teenagers, Gordon Pike and Robin Flewellyn, were standing near the beach's recreation center when they noticed a disturbance in the otherwise calm waters of the lake. Within seconds, a five-foot-tall humanoid-like animal emerged from the water and began running toward them. The young men fled, but not before the creature managed to slash one of them on the hand with razor-sharp spikes protruding from its head. The witnesses described their assailant as having a scaly body, webbed extremities, and a head with "dark, large, bulging fish-like eyes."

Pike and Flewellyn promptly reported the incident to the Royal Canadian Mounted Police, who believed the young men were sincere enough to warrant an investigation. The investigation did not yield any evidence to support their claim, but in due time the

Thetis Lake Monster
(based on a sketch from the Victoria Daily Times, 1972)

creature would return to give the police a second chance. On the afternoon of August 23, Mike Gold and Russell Van Nice, both young boys, were fishing at Thetis Lake when they allegedly saw the creature come out of the water and look around. The location was on the opposite shore from the previous sighting, but the animal's description seemed to match. According to one of the boys:

> It came out of the water and looked around. Then it went back into the water. Then we ran! Its body was silver and shaped like an ordinary body, like a human being body, but it had a monster face, and it was all scaly with a point sticking out of its head and great big ears and horrifying eyes.

The RCMP continued their investigation but failed to reel in the mysterious monster.

Organizations such as The Centre for Fortean Zoology Canada have since looked into the Thetis Lake Monster case and have dredged up some interesting tidbits. First, they examined the theory that the lake monster was actually an escaped pet tegu. Tegus are tropical lizards that can grow up to five feet in length. According to an article posted on the CFZ-Canada blogspot, one of the stories that ran in the 1972 *Victoria Daily Times* included an artist's rendering of the Thetis Lake Monster. This generated a call from a local man who claimed to have lost his pet tegu near the lake one year earlier. The RCMP considered the possibility that the witness had actually seen the large lizard, but of course that doesn't explain the description of an upright amphibian or "lizardman." Nor does it explain how a lizard that requires a tropical environment could survive in the brutal cold of British Columbia.

To cast more uncertainty upon the Thetis Lake Monster case, a writer for *Junior Skeptic* magazine tracked down Russell Van Nice, now 49 years old, who claimed: "It was just a big lie... [Mike Gold was] trying to get attention." Mike Gold was 12 years old at the time he and Nice supposedly saw the creature, so his credibility as a juvenile witness could be questioned. However, even if the second sighting is attributed to the copycat effect, it does not necessarily shoot holes in the original sighting by Pike and Flewellyn. Nor does it explain an incident that occurred much more recently.

It was late summer of 2006 when Jesse Martin decided to try his luck fishing at Thetis Lake. Martin was a professional fishing instructor, working for the Freshwater Fisheries Society of British Columbia at the time, so it was not unusual for him to make the rounds to various lakes near Victoria. However, this was his first visit to Thetis.

Martin arrived in the afternoon and began fishing from the shore. He didn't have a boat at the time, but there were plenty of walking paths around the lake that provided excellent access to the water's edge. Just as he had hoped, he was having great luck so he continued to fish for several hours, reeling in some trophy catches. "I was having a blast exploring this new lake when I came to the realization that daylight was fading fast," Martin stated in his first-hand account posted at the World Fishing Network blog (www.worldfishingnetwork.com). As such, Martin made a few final casts before deciding to call it a day.

Martin then packed his gear and returned to the parking lot where his Ford Mustang was the sole vehicle remaining. "It was completely deserted," he explained. This was nothing new to an experienced urban fisherman, but things started to get spooky when he heard something moving just out of sight in the bushes. He wasn't sure what it was, but it was alarming enough to make him throw his rods into the car and fumble for his keys. As he hurried into the driver's seat and closed the door, he glanced into the rearview mirror. In the hazy twilight, he saw what appeared to be a man-like figure running towards him.

Martin, now in a state of panic, turned the key and started the car. When he glanced back at the mirror, the figure was almost upon him. "As I put [the car] into gear, he made one last lunge for the passenger side door, where his hand smacked against the handle," Martin recounted. "I peeled out of the parking lot and didn't check the rearview until I was on the Malahat highway. I tried to calm myself during the ride home, but I was definitely rattled thinking about what had just happened."

Martin managed to calm down, but when he arrived at home he received one more shock. Before going into his house, he decided to check the area on the passenger side where the thing—whatever it was—had struck.

"As I walked towards it, I stared dumbfounded at what I saw," Martin explained. "Five scratch-marks with patches of fish scales strewn throughout."

For obvious reasons, Martin did not sleep well that night. The next day he was still shaken, so he decided to tell a co-worker of his experience, even if it came out sounding absolutely ridiculous. But his co-worker didn't laugh. After listening to his account, she merely asked him if he had heard of Thetis Lake's strange past. Martin admitted that he had not. So he looked up the Thetis Lake Monster. Martin stated: "... after doing so, I had to admit, it sounded a lot like what I encountered that August night on Vancouver Island. And I haven't been back to Thetis Lake since."

Gatormen

Another group of creatures, which would conceivably share commonalities with "lizardmen," are the so-called "gatormen." Said to possess traits of both alligator and man, these hybrid beasts are rumored to dwell in the marshes and swamplands of the United States.

A prime location for gatorman tales is the Florida Everglades, where whispers of such a creature date back to the 1700s. Though there has been no evidence to show that such creatures actually exist, their profile has been created by such websites as American Monsters (www.americanmonsters.com). According to this site, the creatures are "approximately 5-feet long, with a child-sized torso which tapers off into a long, muscular tail—complete with stubby, gator-like legs." In addition: "Eyewitnesses have described the body of this aquatic-hominid as covered with a thin sheen of greenish scales, with webbing between its toes and fingers and a mouth full of what has been described as 'razor sharp' teeth."

As an example of the archetypal gatorman, the site presents "Jake the Alligator Man," an infamous sideshow gaffe that features a mummified looking head and arms attached to a taxidermied alligator carcass. This clever fake is housed at Marsh's Free Museum in Long Beach, Washington, and was touted in the *Weekly World News* as the "Half-human, half-alligator discovered in Florida swamp."

I spent considerable time trying to locate credible reports of these so-called "gatormen" in Florida, so as to compare them to the Lizard Man sightings, but was unable to come up with any that didn't reek of fancy folklore. My friend and fellow cryptid researcher Ken Gerhard and I were once discussing the subject of gatormen when we thought for a moment we were on to one. He mentioned that a girl he knew once told him that she had seen information regarding a gator-like man when she was younger. However, once we realized she lived in Washington state, we concluded that she must have seen a flyer for good ol' Jake the Alligator man, who is quite well-known in that area.

Even though I didn't uncover any credible gatormen stories from either Washington or Florida, I did find some that originated in the northeastern U.S. These cases, which come from the New York and New Jersey areas, suggest that at least some people have seen something they believe to be a sort of mash-up between a prehistoric reptile and man.

The first report surfaced in 1973. According to Loren Coleman in his book *Curious Encounters*, in the summer of that year residents of New Jersey's Newton-Lafayette area apparently saw what they described as a "giant, man-like alligator." Coleman adds that: "In 1977, New York State Conservation Naturalist Alfred Hulstruck reported that the state's Southern Tier had a scaled, man-like creature (that) appears at dusk from the red, algae-ridden waters to forage among the fern and moss-covered uplands."

Another encounter appears in *HUMCAT: Catalog of Humanoid Reports*. The document states that one evening in the fall of 1974, a man was driving near White Meadow Lake in Rockaway Township, New Jersey, "when his headlights illuminated an immense humanoid figure standing by the side of the road." Though it wasn't facing him, he could tell that it was "greenish in color and covered with scales." As the witness continued to approach, the thing turned its head to the left so he was able to see its profile. He could now see that it was "reptilian, with bulging frog-like eyes and a broad, lipless mouth." The driver, horrified, quickly hit the gas and left creature standing in the darkness.

Scaly Odds and Ends

Outside of the notable, yet brief sightings of the Riverside Reptoid, Loveland Frog, Thetis Lake Monster, and a few scant gatormen, we're left with a smattering of odd reports that make up the rest of the lizardmen case files. Though these too have been presented by Loren Coleman, Mark Hall, John Keel, and others, I will briefly recount them here so we have the complete repertoire of reptilian reports in one place.

The first comes from the extensive files of Loren Coleman. According to Coleman, a man was fishing near Saginaw, Michigan, in 1937 when he witnessed a "man-like monster" climb up a river bank, lean against a tree, and eventually return to the river never to be seen again. No other significant details are provided, save that the man suffered a nervous breakdown following the sighting.

The next incident was investigated by Charlie Raymond, founder of the Kentucky Bigfoot Research Organization. According to his report, in the fall of 1966, a nine-year-old boy in Stephensport, Kentucky, was awakened at approximately 1:00 a.m. when he heard a loud commotion outside his bedroom window. Startled, the boy got up and looked outside. He couldn't see anything moving, so he ran to the living room and pulled back the curtains covering the front door. What he saw would chill him for life. There in the yard stood a brownish-green, man-like creature with scaly skin and webbed hands. "I can only best describe it as a 'lizard-man,' although the only 'human' thing about it was the fact that it stood on two legs and was about 5'6" to 6' tall!" the witness, now 49, told Raymond.

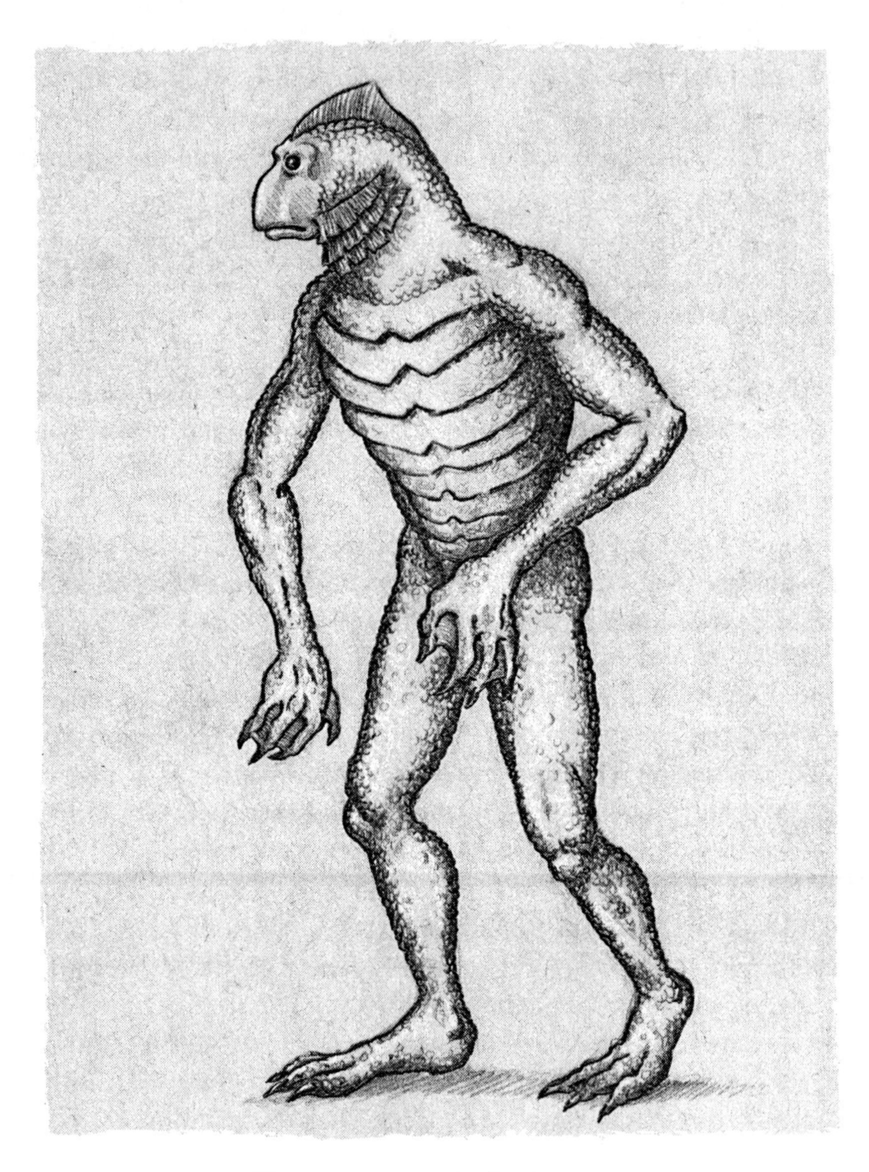

The creature seen in Stephensport
(based on a sketch by Bart Nunnelly)

About the time the boy peered out the window, the creature caught sight of him. The creature immediately took off running on two legs in the direction of a nearby waterway called Sinking Creek. The boy darted to an adjacent window to get another glimpse, but after about 75 yards the thing disappeared in the darkness and could be seen no more.

"It was very 'amphibious looking' with scales covering its entire body," the witness recalled. "What I remember the most about its face were these huge rows of gills which flared out on both sides. Its face was very 'hard' looking with little dark eyes similar to a snake or lizard. I can't recall a nose or lips as the face to face encounter only lasted a second or two. There was also this ridge-like feature which started on the forehead and ran back over the top of its head."

Interestingly enough, Sinking Creek feeds into the Ohio River, along which several of these scaly humanoid sightings—including those of the Loveland Frog—have taken place. The creek is unique in that it's the only natural spring-fed creek in Kentucky. At one point it disappears underground for 12 miles (hence the name Sinking Creek) before it resurfaces in a surge of gurgling water. The creek is also notable since it's believed to be part of the vast underground water system that connects the many caverns of Western Kentucky.

According to Raymond, the witness is now married with two children in college. He seemed very credible and the details of his story had not changed from the original report. When asked if he thought it was possible to have been fooled by a person in a costume, the witness felt certain that wasn't the case. It was too "life-like" and moved too quickly to have been a man, he thought. He also pointed out that this was the remotest part of Breckinridge County. The idea of a person roaming around down there in an elaborate monster suit back in 1966 just seems out of the question.

Another loosely related case comes from the files of John Keel. As detailed in his book *Strange Creatures From Time and Space*, the incident took place in July of 1966. Late one evening, two girls were sitting in a parked car near Lytle Creek outside Fontana, California (a mere 20 miles from the site of Charlie Wetzel's infamous run-in with a reptoid). As the girls were talking (or doing whatever girls do when they park in a wooded area at night), they suddenly noticed a huge creature standing next to the car. Utterly frightened, they sped away from the scene and called the Bernardino County Sheriff's Office. They told police that the creature was at least seven-feet-tall, covered in dark hair, and draped with "moss and slime." No other details are available, but the incident must have caused quite a stir since it triggered a full-on "monster hunt" in which more than 250 hunters poured into the woods looking for the thing. Unfortunately, it was never found.

A reported encounter with a "frogman" occurred in 1952 at Lake Conway, north of Little Rock, Arkansas. According to Mark A. Hall in his book *Lizardmen*, a man by the name of George Dillon was fishing one February morning when he came face-to-face with a strange creature. As he was running his trotlines by boat, one of them became snagged. Thinking it must have been dragged into the brush by a large fish, he began to pry the line with his paddle. When the hooks finally came to the surface, he could see that they had captured some sort of aquatic animal unlike he had ever seen before. "Dillon lifted it partly out of the water, and it began moving its head from side to side, trying to free a hook from its mouth with its tongue," wrote Walter Ed Scales of the *Log Cabin Democrat*, which first published details of the encounter on March 7, 1952.

As the creature struggled, it came forward and put a hand on the boat. "It had green, spotted skin similar to a frog's, and a head which resembles a monkey's, with its temples pinched in, and a monkey-like mouth with blue lips and no teeth," the fisherman said. He estimated its weight to be around 80 pounds and added that it had "fairly broad shoulders, more stooped than a man's" and a humped back. Dillon watched in horrified amazement as the thing freed itself, then pushed off from his boat and swam into some nearby bushes. Once the creature was out of sight, he dropped his trotline and quickly left the area.

Moving over to Wisconsin, we find an interesting account of an alleged "reptile man." As reported by Richard Hendricks on the Weird Wisconsin website (www.weird-wi.com), one day in the mid-1990s a Wisconsin Department of Natural Resources game warden was traveling down Highway 13 near the town of Medford when he noticed a figure standing in the middle of the road. Naturally, the officer slowed down, thinking it was a person in need of help. But when he got closer, he realized it was not a person per se. Instead, what he saw standing before him was a man-like creature with scaly, greenish skin!

The warden pulled his car within several yards of the creature, at which time "wings suddenly popped out from behind the creature's back." The reptilian thing then took flight, passing over the vehicle and landing in the road again behind the warden. It stood there a few moments before eventually flying off into the night, leaving the officer to contemplate just what it was that he had seen.

A flying reptile man is definitely hard to swallow, but this was not the only time this purported creature was seen in the roadways of Wisconsin. According to the man who told the story of the warden, a group of construction workers had also seen a creature fitting the description of a scaly humanoid on the very same stretch of highway. It also had sprouted wings and eventually flew off into the surrounding trees, leaving us to guess at just what kind of creature this might have been, if in fact, it was a real creature.

The remaining reports of bipedal reptiles deal with actual lizard-like animals, which are said to run on two legs instead of four. These creatures do not have any human characteristics but are unique in that they appear to be predominately bipedal in their locomotion. Reports often refer to these animals as "living dinosaurs," but researcher Chad Arment, who has gathered several of these reports along with his partner, Nick Sucik, hesitates to label them as such. Arment explained to me: "We didn't want to call them 'dinosaurs' because there are reptilian possibilities other than 'true' dinosaurs, and it would be silly to force an identification without actual physical evidence."

Sightings of these bipedal reptiles tend to be located up and down the central corridor of the United States. One example would be the "six-foot-tall, dinosaur-like" animal reputed to inhabit portions of West Texas. The creature is said to be "strongly bipedal," but no other details can be found regarding the specifics of these sightings.

Colorado also has a history of reports that suggest some sort of bipedal reptiles inhabit the Pagosa Springs area. A woman there claims that as a child in the 1930s she saw several of these creatures. They were approximately seven-feet-tall with grayish or dark green skin, a snake-like head, short front appendages, stout back legs, and a long tail. She also claimed to have seen one in 1978 when it crossed a field in front of her.

A similar strange reptile was reported near Rawlins, Wyoming. In an interview with Chad Arment the female witness stated:

> [Near] the end of April, 2003, while traveling in Wyoming, my mother and I had quite an experience. A creature of some sort, which looked prehistoric, came down in front of our car. First we saw its feet. They were like huge crocodile or alligator feet. Then we saw his lower body and all of its tail. The tail was scaly... about five feet long, best guess. When it was landing, it came down sideways. Its body was shaped like the pictures you see of dinosaurs. We estimate it was about 7 feet tall. It kept its head in the shadows but it was shaped like a head of a kangaroo.
>
> After sloshing its tail back and forth one time, in a single leap it was gone and out of sight. This happened on I-80 about 40 miles west of Sinclair, Wyoming. I did find a woman truck driver who had seen it and said it was a common sighting there, and people called it the big green dinosaur, even though it was more grayish in color.

Descriptions of these "dinosaur-like" creatures tend to match the pop culture version of the Bishopville Lizard Man—with descriptions of kangaroo-shaped heads and long tails—but in the end do not provide much in the way of helpful comparison. Only the sightings of the more "humanoid" reptiles seem to cast any light on our subject, but even those just seem to raise more questions than answers.

The only thing that is certain is that none of these other cases truly compares to that of the Bishopville Lizard Man, either in the number of sightings, the credibility of witnesses, or the prolonged media buzz. It's surely one reason why the case in Lee County first comes to mind when pondering the possibility of cryptid lizardmen. Whether it's called the Bishopville Lizard Man, the Lizard Man of Scape Ore Swamp, or the Lizard Man of Lee County, there's no doubt that this creature is the seminal "lizardman" of the cryptozoology world.

But just what is this Lizard Man? That's the final and most important question left to answer.

7. Possibilities

I was once investigating a strange sighting that occurred on a county road. A few of my associates and I were at the site interviewing the witness as she took us through the sequence of events; it involved a close call with a large, hair-covered bipedal animal as it ran in front of her car. After a detailed interview, followed by an accurate recreation, we came to the conclusion that the witness was telling the truth. We could not definitively prove what she had seen, but there seemed to be no doubt that she had crossed paths with something unexplainable.

The incident took place on a backroad, which cuts through a rural community. There were several old houses and trailer homes along the route, all amidst thick woods. There was also a small creek nearby, along which several other sightings had been documented over the years. As such, we decided to knock on a few of the closest doors to ask if those residents had seen anything strange. The first resident, an older lady, said she had not, but would certainly let us know if she did. She had heard the reports in the area over the years, but when we told her such a creature had run by within 50 yards of her house, she seemed somewhat disturbed.

Next, we approached a home on the opposite side of the road where we were greeted by a shirtless, middle-aged man who was not the least bit embarrassed that his belly was on display like a great white basketball. Within one second of asking if he had seen a Bigfoot in the area, he confidently informed us that there had been a circus train wreck years ago and that's why people had seen an ape-like creature. I knew better, having previously researched train routes in that particular area, so I politely thanked the man for his time and we went on our way.

Standard explanations like the so-called Circus Train Wreck are commonly cited by skeptics of cryptozoological cases. These are generally blind statements derived from hearsay rather than from any sort of in-depth research or study. I've heard them while investigating Bigfoot-related sightings and, as expected, I heard them while investigating the Lizard Man case. As Cindy and I spoke to locals in the area of Bishopville, some of them were open to the possibility that an unknown creature might be lurking in Scape Ore Swamp and some were not. Some of them, like the man with the basketball belly, happily informed us that a circus train was to blame, while others chose to believe the sightings were the result of mistaken identity, media hype, or hoaxes.

Beyond the standard explanations that perpetually swirl around most cryptid sightings, other, more fantastic theories have been proposed by researchers for these lizardman cases. These are not specific to the Bishopville Lizard Man, rather they are intended to account for any and all scaly or otherwise reptilian humanoids. These theories include rare skin diseases, highly evolved dinosaurs, a race of intelligent subterranean reptoids, and even government conspiracies. Some of these may seem extreme in what many consider to be a purely cryptozoological case, but perhaps this is due in part to the very concept of a "Lizard Man," one that defies the basic rules of biology. It's not a stretch to theorize that Bigfoot (if real) is simply an undocumented hominoid, sharing traits of both ape and man, since it is presumably a creature that could naturally occur in nature. But if the Lizard Man possesses traits of both human and *reptile*, then this becomes more difficult to rationalize. As a result, more bizarre theories have come into play.

But what if the Lizard Man is not reptilian at all? Perhaps, as some have suggested, they are sasquatch that happen to be covered in green slime or mud at the time they are seen. This is much like the mistaken identity theories involving known animals, trees, or other naturally occurring phenomenon, yet with a twist. It may seem ludicrous to explain one cryptid with another cryptid, but it's a final possibility that must also be explored as we attempt to bring ourselves one step closer to solving the mystery of the infamous Bishopville Lizard Man.

DERAILING THE OBVIOUS

The first theory to explore in the Bishopville Lizard Man case is that of the alleged Circus Train Wreck. The Circus Train Wreck, as the name suggests, involves a scenario in which a train or truck carrying circus animals crashes, thereby releasing non-indigenous animals into the local woods. The result is a situation in which witnesses believe they have seen an unknown creature when in fact they have seen something unexpected but still wholly explainable.

While admittedly there aren't too many folks who claim that the Bishopville Lizard Man is the result of a crashed circus transport, at least a few have brought it up. However, there are two main issues with this scenario: (1) There needs to be an actual circus train wreck in the area for this to be possible, and (2) There would have to be an escaped animal that fits the description of the Lizard Man for it to be mistaken as such.

In searching the various newspapers in the area, I could find no evidence that a circus train (or truck) crashed anywhere near Bishopville. Circus train crashes typically make sensational and tragic news stories, so it's fairly safe to assume that if one had crashed in central South Carolina, then the news media would have picked up on it. I did find stories of crashes in neighboring states, such as Georgia, but these were too far away for even the heartiest of circus animals to hitchhike all the way to Scape Ore Swamp.

Since there are no circus train mishaps on record near Bishopville, this essentially derails the possibility that any such event could have contributed to the Lizard Man phenomenon. But even if I've overlooked a past incident, we know that circus animals typically include elephants, camels, bears, big cats, and chimpanzees. Since no one has dubbed the Bishopville creature as the "Elephant Man," "Camel Man," or "Tiger Man," we can pretty much toss those out. Bears and chimps are somewhat closer to some descriptions of the Lizard Man, but only by a stretch of the imagination. These creatures primarily move on four legs, which is not what was reported.

Black bears are not exclusive to passing circus troupes, of course, and have been reported in Scape Ore Swamp, although again, the majority of Lizard Man descriptions don't seem consistent with a bear. A few witnesses, such as the Blythers family and the Elmores, did report seeing hair on the creature, but stressed that it *wasn't* a bear. As Michelle Elmore stated in her police report: "I have seen bears and gorillas in circuses and the zoo. It was defiantly [sic] neither." The only incident in which a bear might have been the culprit is the sighting by Rodney Nolf and Shane Stokes. The boys claimed to have seen a large, muscular creature run across the road, which, some could argue, was actually a bear.

Another zoo-type animal that made the list of Lizard Man suspects is the peacock. According to a 1991 article in the *Spartanburg Herald-Journal*, some locals theorized that the "Lizard Man was really a green peacock." Lee County residents Johnny Brown and Clyburn Davis told reporters that they had seen a "green peafowl ambling around the woodlands near Browntown," which caused them a second look. "I thought sure he was the Lizard Man at first," said Brown. "But he's a great ol' big green peacock." The creature was primarily green in color, but very few residents put stock in the idea that a two-foot-tall bird of any sort could be mistaken for a seven-foot-tall upright, humanoid creature. "You'd have to be more than a little drunk to confuse a Java cock with Lizard Man," laughed Bob Seibels, curator of birds at the local Riverbanks Zoo. Truesdale scoffed at the idea as well.

A related theory deals with the possibility that an exotic pet may have escaped from a private owner and fled into the bottomlands of South Carolina. But again, this does not provide a plausible explanation. People tend to own big cats, monkeys, chimpanzees, birds, and small reptiles, none of which match descriptions of the Lizard Man. Yes, some people do own lizards, and even alligators, but these are not the seven-foot-tall bipedal sort.

A final prospect related to circuses, or more specifically sideshows and carnivals, is that the creature is not an animal, but a human with an animal-like appearance. People with a genetic skin disorder known as *ichthyosis* can look reptilian or fish-like since the condition causes their skin to be exceptionally thick, cracked, and scaly. (The very term *ichthyosis* is derived from the Ancient Greek *ichthys*, meaning "fish.") Although the severity of deformity varies from case to case, some instances are fatal, and some are horrifying enough to cause the person to be shunned from society. As a result, many victims in times past ended up in sideshows, billed by such monikers as "fish boy" or "fish man."

The condition does present an interesting possibility, but not a likely one. If an adult with an extreme case of ichthyosis were to be seen creeping around in a swamp at night or charging across a moonlit field, it would certainly be startling but probably not so much that it would not be recognizable as mere human. As well, the condition does not result in green skin, glowing red eyes, three fingered hands, or any other monstrous traits described in the various Lizard Man sightings. Not to mention that it seems preposterous that a person with ichthyosis would be out running around naked (nobody reported clothes!) in the first place. In the end, I think we can assume that the Lizard Man must be something other than a known animal or a person with ichthyosis.

MAN-MADE MONSTER

Another theory, and one that we touched on previously, is that the Lizard Man can be attributed to a person or persons, namely Brother Elmore, the farmer who was trying to scare people away from his Butterbean Shed. This presumption, which is often brought up when discussing the Lizard Man case, is widely held by locals, who either knew Elmore or have heard the various rumors over the years. "Brother Elmore is the Lizard Man," they exclaim!

Without interviewing the late Brother Elmore myself, I can only make reasonable assumptions based on the entire body of evidence found in the witness statements, press articles, and my interviews with others. This may not be the best way to get to the truth, but regardless it seems nearly impossible that a man could have been responsible for all of the sightings.

As we learned, the main incident attributed to Elmore is the Chris Davis encounter. If Davis is telling the truth, it's hard to reconcile how Elmore could have performed all the feats that the Davis creature accomplished, including running 35 mph and jumping on the roof of the car. But suppose that Davis was lying, and it was actually a burlap-clad farmer who ran him off that night. Elmore would still have to fool at least 11 other witnesses over a span of more than five years in order to earn the title of Lizard Man.

Let's review the witness list:

1) George Plyler
2) George Holloman
3) Frank Mitchell
4) Chris Davis
5) Rodney Nolf and Shane Stokes
6) Colonel Mason Phillips
7) Bertha, Johnny, and Tamacia Blythers
8) Brian and Michelle Elmore

Now let's discuss each sighting to decide if Brother Elmore or any other person could have been responsible:

George Plyler: Mr. Plyler described an odd creature with some human-like traits but one that did not seem like an actual human. Among his descriptions, Plyler said that "its face was shaped like a human's, its eyes were red around the pupils, and its arms hung like an ape's." I suppose it's not out of the question that it could be a human dressed in a costume, but why would someone dress up and creep around a swamp to scare a local man and a few workers? Not only that, Plyler said this was several years prior to the Davis sighting. All in all, it makes very little sense that a person would be behind it.

George Holloman: Holloman's sighting occurred in a wooded area at night, so naturally his view of the creature was not crystal clear. Nonetheless he believed it to be "7 to 8 feet tall" and covered in hair. The height tends to rule out a human, but even if it were a human dressed in a suit, what in the world would a person be doing in Browntown dressed like a Bigfoot? It certainly doesn't seem like something a farmer or anyone else would have reason to do.

Frank Mitchell: Mitchell's sighting is a huge strike against the Brother Elmore theory. Not only did Mitchell claim to see a hair-covered creature that didn't appear to be human, he claimed to have seen it at the same time Brother Elmore was standing in his field. This doesn't rule out an accomplice dressed up in a monkey suit, but again, why?

Rodney Nolf and **Shane Stokes**: These two boys described a large, muscular creature, which crossed in front of their car headlights. This definitely doesn't seem like the physiology of a human, nor does it seem like the actions of a human. If it was not an unknown creature, then perhaps it could have been a bear, but either way it's highly unlikely that it was the work of Elmore or another person.

Colonel Mason Phillips: The Colonel claimed to have seen a large, brownish creature running on two legs in front of his car. Much like the sighting by Nolf and Stokes, it appeared to be some sort of strange animal, not a human. The only puzzling detail of this sighting is that he claimed to see a tail, whereas no one else had reported seeing one. It's not beyond the realm of possibility that someone had fabricated a suit to fool passersby, but why would they not do their hoaxing closer to the infamous Scape Ore Swamp bridge? Phillips claimed the sighting took place several miles away, which does not rule out a hoax, but does cast considerable doubt that it was Brother Elmore or an accomplice doing the deed.

The Blythers: The creature described by the Blythers most closely resembles a sasquatch. If it were a person, they would have had to create a convincing suit and lunge at a passing car in the dark to pull off the stunt. If this were

Elmore or even some kids trying to scare people near Scape Ore Swamp, then it seems strange that they would not have created a costume that more closely resembled the public's perception of the Lizard Man. That fact that this was a large, hairy ape-like creature—sighted two years after the Davis incident—leads us in a direction which is both away from scaly humanoids and away from a farmer posing as a monster.

The Elmores: While they do share the same last name as Brother Elmore, Brian and Michelle Elmore were only distantly related to the farmer, so they were most likely not a party to any proposed hoaxing scenario. Their sighting was nearly identical to that of the Blythers, so again, it seems unlikely that this was the work of Brother Elmore.

After examining these incidents, it doesn't seem possible that Brother Elmore alone could have been responsible. Pulling off a hoax would have required the help of multiple individuals working over the span of several years, which seems highly unlikely. There is, however, one incident outside of the eyewitness accounts that is unquestionably a hoax… the famous three-toed tracks discovered by the police in 1988.

When I examined the track casting of this famous footprint at the Cotton Museum, my first impression was that it could not have been made by a real animal. The track is just too oddly constructed and almost too "monstrously perfect." The two pads that make up the body of the foot are very well defined with the front pad being nearly identical to the rear pad. The precise definition does not ring true with typical animal tracks. When the sole of the foot, made up of flesh, bone, and fat impacts the soil, very rarely does this result in such well-defined, 90 degree shapes as we see in this case. The fact that the rear pad is as large as the front pad is also a red flag. In typical foot anatomy, whether the subject is bipedal or quadrupedal, the heel of the foot is generally smaller than the forefront. The evidence shown in the Bishopville casting not only opposes logical anatomy, but it also appears to have been manufactured using the very same oval object to create both the front and rear pads.

The toe formation likewise poses problems. The fact that the alleged bipedal creature has only three toes notwithstanding, the lengthy finger-like extrusions, like the pads, seem altogether unnatural and "manufactured." They are all the same size and shape suggesting that the same object was used to create each one.

Evidence of a hoax becomes most clear when the various casting samples are compared to each other. When scanned and overlaid, it turns out they are identical, which is the ultimate red flag. Naturally occurring footprints will vary from imprint to imprint as the animal walks due to subtle changes in the ground, the force of impact, and the very nature of flesh. Only footprints that have been manufactured using a single template could be so identical.

All of these factors undoubtedly came to the attention of the wildlife officers who initially examined the tracks and concluded that they could *not* have been made by a real animal. This was a polite way of saying it was "hoaxed." In fact, several credible locals told me during my investigation that they knew the tracks were hoaxed and could name the suspect. This was also reinforced by a conversation I had with Josh Gates, star of SyFy's *Destination Truth,* who had looked into the Lizard Man case as part of his television show. Gates told me that while he was in Bishopville filming for the show, an unnamed local also admitted to him that the tracks were a fake. As part of the show, he and his team clearly demonstrated that the track castings were identical.

Knowing the tracks are fake, what does this do for our overall Lizard Man case? It certainly doesn't bode well for any case when there is hoaxing involved, but in reality, this particular bit of evidence only served to confuse matters since the footprints are so ridiculous and unnatural. Throwing this out of court eliminates the worry over what kind of animal could have possibly made such bizarre tracks. The tracks merely illustrate a weak attempt at a hoax that might have fooled a few people in the moment but do not stand up to the ample eyewitness accounts that firmly suggest something truly unexplainable once visited Bishopville.

Dinosauroid

If the Lizard Man is not a known animal or a hoax, then what are the other options? Given the perceived reptilian traits, some have suggested the creature is some sort of highly evolved dinosaur. It's hard to imagine a dinosaur with such an anthropomorphic body, but there is one controversial hypothesis that suggests it could have been possible, if certain dinosaurs had continued to evolve.

Known as the "Hypothetical Dinosauroid," details of this conjectured evolutionary path were conceived by D.A. Russell and R. Séguin and published in 1982 in the National Museum of Canada's *Syllogeous* No. 37. The paper first examines a small, cretaceous theropod dinosaur known a *Stenoychosarus ineqalis* (similar to the infamous veloceraptor), noted for its "large brain, stereoscopic vision, opposable fingers and bipedal stature." The paleontologists then discuss their process for creating a life-size, color model of the species based on the most complete specimen of a *Stenoychosarus*, which was found in Alberta, Canada. The paper also goes on to speculate "how descendants of S. *ineqalis*… might have changed had they survived the Mesozoic extinctions, and achieved an encephalization quotient [i.e., brain mass] similar to that of *Homo sapiens*." According to Russell and Séguin, a parallel can be drawn between the brain development of *Homo sapiens* and that of the *Stenoychosarus*:

> A curve representing the maximum level of encephalization known in organisms living on our planet can be drawn across the last 600 million years of Phanerozoic time...Man and his immediate antecedents lie on this curve, and *Stenoychosarus* closely approached it during late Mesozoic time.

Without getting too bogged down in scientific jargon, Russell and Séguin essentially hypothesize that if the brain of this dinosaur had continued to grow in proportion to its body, then evolutionary changes would have been necessary to accommodate the new cranial mass. One of these changes would have been a more upright stance, since the "tendency to position the head more directly over the vertebral column is seen in anthropoids of increasing brain size." This would conceivably lead to more anthropomorphic legs, longer arms, and most dramatically, a human-like head.

As with the *Stenoychosarus*, Russell created a realistic "life-size" model to represent the dinosauroid. The result is a striking form that looks something like a "lizard man." Its skin is green; its eyes are large and bulbous; it has a toothless, beak-like mouth; it has three fingers on each hand; and it stands upright on two legs. Coincidentally (or perhaps not), the dinosauroid bears a striking resemblance to the sleestaks from *Land of the Lost*.

The dinosauroid is an intriguing thought exercise but one that has been met with plenty of criticism. Many paleontologists argue that the creature is overly anthropomorphic, in other words, too human. On his Tetrapod Zoology blog, paleontologist Darrian Naish sums up the sentiment with this statement:

> ... my feeling on dinosauroids and intelligent theropods and so on is that—if they were to evolve—they wouldn't look like scaly, or feathery, people, but would instead be far more normal from the theropod point of view. A horizontal body posture, not a vertical one. Digitigrade feet, not plantigrade ones. A long tail, not a reduced one.

But regardless of whether tetrapods would have taken an evolutionary path similar to that of humans, they were (as far as we know) never given the opportunity. The same mass extinction that wiped out the majority of dinosaurs would have also ended any opportunity for them to develop beyond their initial body form. It's true that some smaller reptiles developed into modern day birds, but these are far from human in form. So even though the dinosauroid would be a near perfect fit for our Lizard Man, the likelihood that a small population of tetrapods managed to survive extinction, evolve into human-like creatures, and remain hidden in

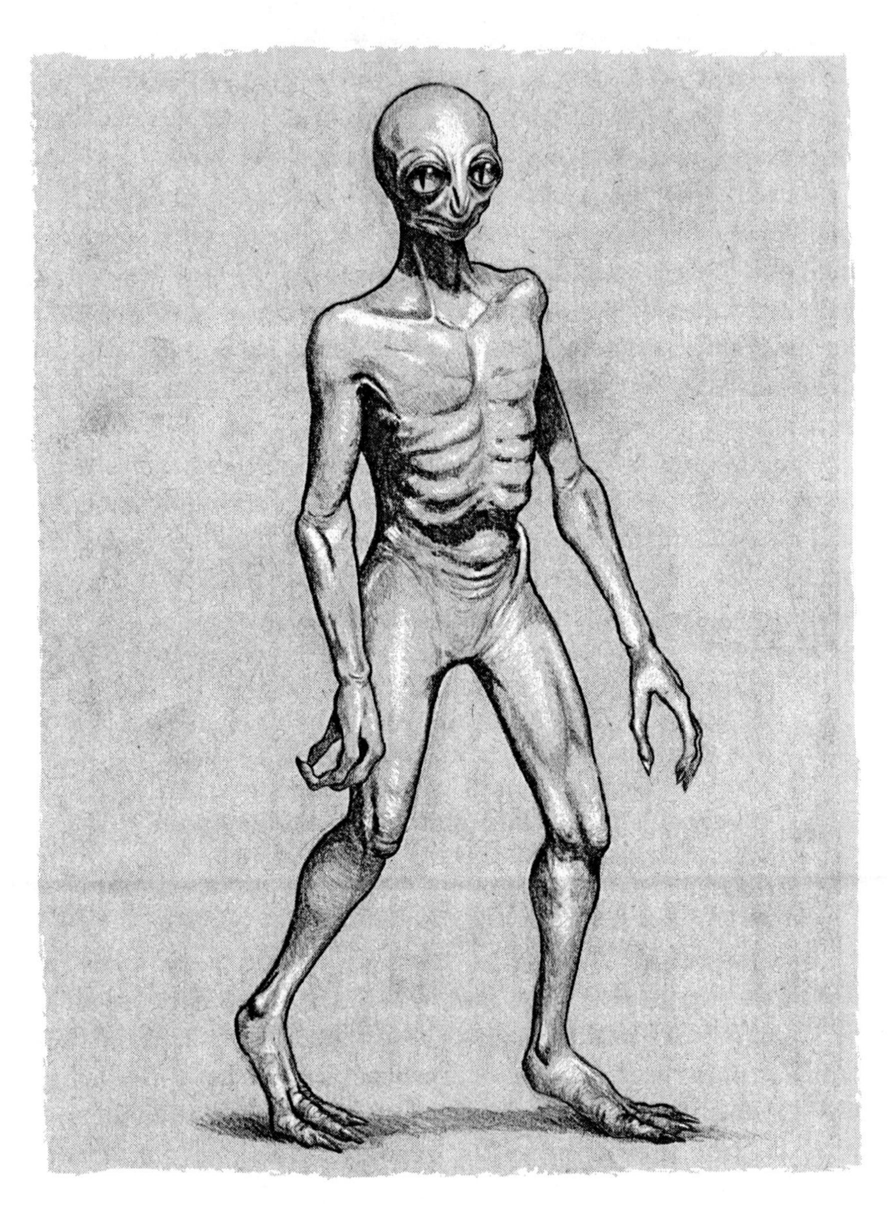

Russell's Hypothetical Dinosauroid

the wilds of South Carolina is all but impossible… or is it? Some may argue that the creature seen in Scape Ore Swamp is but one of many highly evolved reptiles that do indeed walk among us.

REPTILIAN AGENDA

One of the most radical theories involving lizardmen is that of the "underground reptoids" or "reptilian agenda." Supported by an array of researchers, speakers, and writers, the theory states that a race of highly evolved and highly intelligent reptilian creatures inhabit a vast system of underground caverns. These creatures, believed to be descendents of an ancient race of "reptoids," are not only thriving in this unexplored world but are secretly interacting with select humans for the purpose of exchanging scientific and military knowledge. This theory will no doubt leave some people scratching their heads, but it doesn't stop there. Some even claim that these reptilians are able to shapeshift into human form, making it possible to walk among us undetected. They have allegedly sought out positions of authority with the intention of manipulating us through politics or other means. It's like the classic John Carpenter movie, *They Live*, with a twist. The unseen invaders aren't from outer space; they have been here all along, literally beneath our feet!

The idea of shape-shifting reptiles obviously has the potential to lead us off on wild tangents, so we will explore this concept only as it relates to the Bishopville Lizard Man case. In other words, was the creature sighted in Scape Ore Swamp part of a larger population of intelligent underground dwellers?

To understand the premise of this concept, it's best to consult the website of John Rhodes. Rhodes is one of the foremost researchers seeking to prove that we share the planet with underground-dwelling reptoids. His "Terrestrial Reptoid Hypothesis" states that "the beings that humanity has been calling 'E.T.'s' or 'Aliens' are (in many cases) sightings of sentient Reptilian-Humanoid 'Reptoid' beings that are native to Earth, or fellow 'Terrans.'" It further asserts that "the ancestors of the reptoids (and possibly other life forms) retreated into underground caverns during environmental cataclysms long ago and there they continued to thrive and evolve into the sentient beings they appear to be today." According to Rhodes, these beings can exist in three main places: under the earth's surface, in space, and in alternate dimensions. They can also, on occasion, be found "in remote (and sometimes local) geological locations (such as state parks and wilderness areas)." Rhodes includes a page on his website with information about the Bishopville Lizard Man, and although he doesn't explicitly state it, one can only assume that he believes the creature to be one these Terrestrial Reptoids.

Another proponent of an underground reptoid theory is David Icke. Icke contends that not only do intelligent reptilian beings live in caverns below us, but that they have the ability to shapeshift. Icke's radical theories—disseminated by his books, website, and speaking engagements—have undoubtedly made him a controversial figure.

Icke first introduced his reptoid hypothesis in *The Biggest Secret*, published in 1999. His assertion is that a group called "The Brotherhood" were descendants of reptilians that came from the constellation Draco. They appear human but are actually 12-foot lizards that live in subterranean caverns within the earth's core. Over time, their descendents have been able to infiltrate the world's governments to carry out an agenda to control mankind. Among others, Icke has accused politicians such as John F. Kennedy, George W. Bush, and Al Gore as being part of this "reptilian bloodline." He's also pointed a finger at other famous people from royalty to musicians, including Queen Elizabeth II, Kris Kristopherson, Boxcar Willie, and even Bob Hope.

As ridiculous as this may sound, the point here is that some people believe a race of reptilian humanoids share the earth with us. On occasion, it is said, they venture to the surface world where they are seen by humans thus resulting in cases like the Bishopville Lizard Man or the Riverside Reptoid. Whereas some consider these creatures to be strictly cryptids (undiscovered animals), others, like Rhodes, believe them to be part of an organized, intelligent race.

While there's no way to prove or disprove the theories of Terrestrial

Reptoids or the Reptilian Agenda, it is curious that native Americans in the South Carolina area told tales of the lizard-like men from Inzignanin. Was Inzignanin, in actuality, an underground cavern instead of a location on the surface? It's certainly a mind-bending notion, although it can be said with some amount of certainty that there are no caves or networks of subterranean caverns near the Scape Ore Swamp. (South Carolina has very few significant cave systems, with the closest being the Santee Caves located in Orangeburg County some 70 miles away.) And if there are no caves, then how would an underground-dwelling creature emerge in that area? Where is the entrance and exit to such a place? As well, if these beings are intelligent, then why chew up car bumpers, attack a teenager, or hang around the Scape Ore bridge and frighten a guy on a bicycle? What could possibly be gained by this?

If we take it one step further and assume the Bishopville Lizard Man is one of these powerful, shape-shifting reptoids, then why in the heck would it be hanging around in a muggy swamp in South Carolina? Should it not be rubbing scaly elbows with politicians down at the governor's mansion? Sightings of a seven-foot-tall scaly humanoid creatures are hard enough to swallow, so it only becomes more confusing when trying to tie them to a larger, more complex scenario. It's not to say that aspects of the reptoid theory aren't possible—the world is full of mysteries we don't yet understand—only that it seems improbable that the creature seen near Bishopville is a highly intelligent terrestrial from an underground city.

NOT WHAT IT SEEMS

One of the most promising scenarios is that the Bishopville Lizard Man is not what its name suggests. In other words, the creature is not actually reptilian but a more familiar primate-like cryptid. Yes, Bigfoot. Given the amount of Lizard Man witnesses that reported seeing hair or other ape-like characteristics, it does seem like a reasonable explanation.

Let's review once more the traits described by the witnesses:

George Plyler: Mr. Plyler noted that "its arms hung like an *ape's*."

George Holloman: Holloman described the creature as being more than seven-feet-tall and *covered in hair*.

Frank Mitchell: Mitchell said he saw a "grayish-brown thing with a face like a *monkey*."

Chris Davis: Davis said the creature was "green, wetlike, about seven feet tall and had three fingers."

Rodney Nolf and Shane Stokes: The boys described a generic "*large, muscular* creature."

Colonel Mason Phillips: Phillips reported seeing a large, brownish creature with a tail running on two legs.

The Blythers: The Blythers described a bulky creature cover in *hair* which moved on two legs.

The Elmores: Brian and Michelle Elmore flat-out said "it was a *Sasquatch*."

In at least five out of the eight encounters, the witnesses described the creature as potentially being a Bigfoot. In one of the encounters, the witness could not see the creature well enough to identify hair, although it was still described in terms of "large" and "muscular." Only in the Davis and Phillips sightings did the witnesses use descriptive words that might imply reptilian traits, such as "green," "wetlike," and "tail." So are there two Lizard Men or is the Lizard Man merely a Bigfoot masquerading under a misnomer?

To evaluate this theory, I decided to research Bigfoot sightings in South Carolina. If these alleged creatures have been reported in other parts of the state, then perhaps they could, on occasion, pass through Scape Ore Swamp. To start, I accessed the website of the Bigfoot Field Research Organization (BFRO). I discovered that their database contained a total of 51 reports from South Carolina (at the time of this writing). This is not an exceptionally high number, but it's enough to account for 1.15% of all Bigfoot sightings in the U.S. and give it a rank of 29 out of 49 U.S. states. (Hawaii is not included, while Washington comes in at the number one spot with 567 sightings in the database.) Out of the 51 reports, however, I found that the eight incidents filed for Lee County all had to do with the Lizard Man, mostly coming from various newspaper reports (which we have already covered). So that leaves 43 total Bigfoot reports in South Carolina that are not Lizard Man related.

The BFRO is not the only organization to gather Bigfoot reports, of course, so it seemed probable that additional reports had been collected by other, less-publicized organizations or individual researchers. To find out, I contacted Mike Richberg, a cryptid researcher living in South Carolina. In speaking with Richberg, I learned that not only has he personally collected more than 120 Bigfoot reports statewide, he has also seen hairy, man-like creatures himself! The first time was at the age of fourteen. It was an event that would change his life... forever.

It was well before sunrise on a fall day in 1978 when Mike and his father headed out to meet a group of fellow deer hunters southeast of Columbia. They were joining them for a group hunt along the Congaree River at a place known as the Devil's Orchard Swamp. The plan was to use trained dogs to flush out the deer and send them in the direction of the hunters who would stake out positions along one side of the river. It seemed like a sure-fire plan and an excellent opportunity for Mike to bring down a big buck. He was nervous with excitement.

After a quick rendezvous and breakfast at a local diner, the hunters piled into a couple of trucks and headed towards the Devil's Orchard. Mike and his father had been assigned the southern-most position along the Congaree—which flows around the swamp basin—so they were last to be dropped off. The sun was just beginning to light the eastern sky as they headed into the woods on foot.

Once they got close to the river, Mike's father told him to head down to the bank and find a place to wait for the deer. His father then set out to find his own position upstream. Mike did as instructed, jumping over a small stream and making his way through pockets of fog and thick trees until he was almost to the bank that overhung the water. At that point, he found a comfortable place to sit on the edge of a small clearing.

"I was sitting there for a few minutes when I heard something get up out of the river and climb up the bank, just over my left shoulder," Richberg told me during one of our conversations. "I was thinking 'deer,' so I shouldered my shotgun and took the safety off." But as the animal got closer, he got the feeling it wasn't a deer. "That thing rustled around in the brush a little bit and it sounded real heavy," he continued. "So I got to thinking it was a big hog or something, which wouldn't be good since I had a little ol' 20-gauge."

Despite his concerns, Mike decided to remain still and wait it out. Perhaps the animal would reveal itself or would move off in another direction without confrontation. However, the situation grew more alarming when it made a strange sound.

"The best way I can describe it is like a sigh or a grunt... a very human-like sound," he explained. "And that really threw me off, so I got to thinking maybe it was some*body*. Being fourteen, I started thinking all kinds of crazy stuff like a bank robber, escaped convict, or something like that."

The thought of an escaped convict creeping around in the brush was unsettling, but what he saw next was altogether disturbing. As Mike watched, the culprit finally stepped into sight approximately 40 feet away. It had its back to him, but even so he could tell that it was no ordinary animal. This creature was covered in black hair, like that of a bear, yet it did not have the rounded ears of a bear and stood on two feet like a man. It was nearly seven feet in height with huge, muscular arms and shoulders. It was still dripping wet, as if it had walked right up out of the river.

"My first thought was *bear*, but it looked more like a *gorilla*" Mike recalled. However, he knew it was not a gorilla. No gorilla could look so much like a man, yet not be a man. "I didn't know what to think. It just scared me because the thing was tremendous.

A hairy, man-like beast is seen on the banks of the Congaree River

It was huge."

Mike held his breath as he watched the creature walk to the place where he himself had entered the clearing. For a moment it stopped and turned its head to the left as if it had picked up a scent. Mike figured surely it had winded him, but after a few moments it continued walking into the trees, pushing aside branches with its hands as it went. Mike let the air out of his lungs as he began to shake. He watched the creature for several more seconds until it was completely obscured by the brush. Eventually the sound of its footsteps faded and it was gone.

By now the teenager had tears streaming down his cheeks. The ordeal of seeing such an unbelievable creature had overwhelmed him. Sure he had a gun, but not knowing exactly what he was looking at, he didn't dare shoot.

Mike waited a few more minutes before switching on the gun's safety and running for the two-rut road where he and his father had been dropped off. When his father finally came back, he could tell his son was upset, but Mike played it off like he wasn't feeling well. He couldn't bring himself to tell his father what he had seen. The fear of ridicule was too great, so he remained silent. It was an incident that would both haunt his dreams and inspire him to become a serious cryptid researcher later in life.

Given Mike's testimony, it seems promising that Bigfoot creatures may be traversing the areas near Scape Ore Swamp. Devil's Orchard is 50 miles away—as the crow flies—but still within the range of a large, mobile animal. And this was not Mike's only encounter. After years of being afraid of the woods, and finally overcoming that fear, he was able to get a glimpse of another one of these massive creatures at a location 30 miles from his first sighting.

It was in the late spring of 2012. Mike was on a daytime hike through the woods of Sparkleberry Swamp, near the confluence of the Congaree and Wateree rivers, where he planned to set up a game camera. On a few occasions since the 1978 incident he had seen shadowy figures moving through the woods at night, but he could never be certain these were the same type of creatures, or even unknown creatures at all. Despite this uncertainty, he was still optimistic he could capture one on camera.

As Mike hiked toward his destination, he was forced to cross a creek. The bank was several feet high, so he descended down and began to wade across the water. As he approached the opposite bank, he heard two animals move above him. The vegetation was such that he could not see the animal closest to him, but he could see the other one as it suddenly bolted away at an incredible speed. The creature was covered in dark hair and appeared massive and ape-like, just like the creature he had seen back in 1978. It was running quadrupedally at a distance of approximately 80 yards from the creek.

"It was running on all fours in a real funny manner because its arms were so big," Mike explained. "Its chest sort of bobbed up a couple of times, then on the third bob, it went bipedal." Mike was shocked. The animal was running at such a high rate of speed, it would have easily run down a deer. Yet this thing was switching from a four legged stride to two legs in one smooth motion. "It blew my mind that something that big could move that fast," Mike said with obvious amazement.

Mike watched for a few fleeting seconds until the animal disappeared into a copse of thick trees. He was packing an 8mm camera in addition to his game camera, but because he was crossing the creek at the time of the incident and the speed at which it all happened, he was unable to get the elusive creature on film.

Richbergs's personal sightings, combined with the hundred plus reports he's received over the years, have convinced him that some sort of ape-like creatures are living in the woods of South Carolina. According to his data, there are perhaps two different kinds present. The first is described as being at least seven-foot-tall with reddish-brown hair and human-like features. These animals are generally seen in the Pee Dee region of the state, an area which encompasses the northeastern counties, including Lee. The second creature is described as being shorter—always under seven feet—with black hair and a more gorilla-like body. These animals, which include those sighted by Richberg himself, are generally seen in the low country Coastal region near the Wateree, Congaree, and Santee

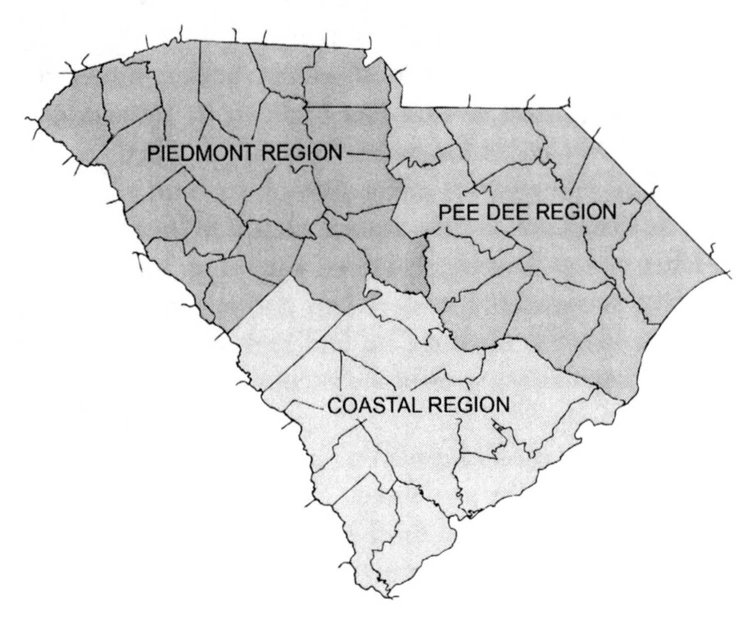

South Carolina Geographical Regions

rivers.

Could one of these animals be responsible for the Lizard Man sightings near Bishopville? It's certainly possible. The only glitches are the two occasions when the creature was described in reptilian terms, primarily in the case of Davis who said it was wet, green, and had three fingers. Perhaps this can be accounted for by Coleman's theory that hair-covered, Bigfoot-like creatures can occasionally have "patchy hair growths that appear 'like leaves' or 'scaly.'" If a Bigfoot with this condition were to live in a watery bottomland area and be covered in algae-rich mud or moss, this could explain its green, wet-like appearance. It doesn't explain the three fingers, but greenish mud which has dried and cracked could certainly give a scaly appearance.

To account for the three-digit hands described by Davis (and associated three-toed feet), some have proposed that the Bishopville creature is deformed due to inbreeding within a small population. Others purport that it's one of a specialized breed of "water ape." In his book, *Lizardmen*, Mark A. Hall theorizes that a surviving breed of ancient primate could have evolved into an aquatic or semi-aquatic form. He cites fossil records of the *Oreopithecus*, a "swamp ape" known to have existed around nine million years ago as a possible candidate. "One of the most complete fossils that turned up in 1958 was found in what appeared to be a position of swimming," he writes.

This is not to suggest that these apes were aquatic at the time, but perhaps they managed to survive and evolve into something closer to the "lizardmen" seen today. "What seems possible is that over the time of 9 millions of years that *Oreopithecus* was finding its way in the world of inland bodies of water and in marine environments, it branched into diverse forms of water-dwelling primates," Hall concludes.

It's also possible that the creature encountered by Davis was not the same type seen by the other witnesses. This could account for its green, scaly appearance and also for the three-fingered hands, which obviously throws a wrench in the theory that it was merely an unknown primate. If Davis did in fact see an unknown creature, then perhaps a third human-like monstrosity is roaming the bottomlands of South Carolina… one that has three fingers and three toes.

The famous Lizard Man tracks don't help the case for Bigfoot, however, as George Plyler claimed to have found "three-pronged" tracks in the area where he saw the creature. As well, Mike Richberg stumbled on a three-toed trackway that he felt was legitimate. It was around 1989 after the Lizardmania had died down. Richberg, wanting to research the area for himself, managed to locate a farmer living on the edge of Scape Ore Swamp who was willing to let him explore his property. For weeks he scouted the area, but saw nothing of interest. Finally one day he was walking along the edge of a field near the swamp when he spotted a line of strange footprints. The prints were unusual in that they were long and narrow with three toes. A few of the tracks suggested that perhaps a fourth toe or dew claw may have been present, but he could not be certain. Richberg measured the impressions, which were on average 13½ inches in length by 4 inches wide. The trackway itself covered a distance of nearly 25 yards across the field before it veered into the woods.

Richberg was stunned. Most Bigfoot reports in the state described a creature with a five-toed track, so these certainly presented a genuine mystery. However, three-toed feet are often associated with swamp monsters, especially those said to live in the southern United States. The Fouke Monster of *The Legend of Boggy Creek* fame is one such creature; the Honey Island Swamp Monster of Louisiana is another. The skunk apes of Florida are also said to have three toes. In addition, scaly humanoids, such as the Thetis Lake Monster, are depicted as having three toes. This stereotype includes the Bishopville Lizard Man, although this almost entirely relies on the tracks from 1988.

So could these previously undocumented tracks discovered by Richberg be the true footprints of the Lizard Man? Is this evidence that there was some other creature stalking the banks of Scape Ore Swamp besides a Bigfoot? The theories we have explored can only offer possibilities, but perhaps these are enough to draw our own conclusion as to what may—or may not—exist in the wilds near Bishopville.

8. Conclusion

Our research trip was coming to a close. Cindy and I had spent nearly a week visiting locations, digging through documents, and interviewing people in our quest to bring forth the full story of the legendary Lizard Man. I would still have much more work to do back at home—sorting through more old newspapers, researching the related scaly humanoid sightings, and fleshing out the various theories—but much of the story was already clear. It was a complex and mysterious case, but one that had gripped the public, and myself, with its monstrous, three-fingered claws.

Since it was our final day in Bishopville, Cindy and I wanted to make the most of it. We planned to drop by Truesdale's one last time, then head over to Scape Ore Swamp for some exploration and hiking. As such, we packed our things and said goodbye to the musty motel room that had been our home for the week. I would miss the quaint hospitality of Bishopville, but not so much the lumpy pillows and stale air representative of our economy level stay.

After stopping off for some coffee and eggs, Cindy and I arrived at Truesdale's house. Unlike the first time we pulled up in his driveway, the sun cast a cheerful light across the neighborhood. It was altogether friendly and full of Southern charm.

But we weren't here to document the ways in which a Bishopville morning could be charming; we were here to talk to Truesdale about a rather melancholy epilogue to the case. A few days before, he had mentioned something about some of the Lizard Man witnesses dying under tragic circumstances. Unfortunately, we had gotten side-tracked and had yet to finish that conversation. Since then I had been wondering… was there a Curse of the Lizard Man?

Curse of the Lizard Man

On June 17, 2009, Christopher Davis was at home with this girlfriend, his brother, his 12-year-old daughter, and 14-year-old son, and a friend. By now Davis was 37 years old and living in Sumter, South Carolina, a short distance from Bishopville. It had been more than 20 years since he had become an apparent victim of the Lizard Man, but on this night, he would become victim to something far worse. At around 11:00 p.m., two gunmen broke into the home and searched for Davis. When they found him, they delivered a permanent and deadly message from a shotgun. The men then fled the home, leaving the others unharmed.

According to an article from the June 20, 2009 edition of *The Item*, Davis died instantly in what they believed was a "drug-related incident." Further investigation confirmed that narcotics officers "recovered 10 grams of marijuana and two scales from the kitchen."

Police could not identify the shooters at first, but in short order, a 19-year-old boy turned himself in to Sumter City-County Law Enforcement officials. He confessed that he and a 22-year-old friend had been the perpetrators of the violent crime. Naturally, a manhunt for the second suspect was launched. The effort resulted in his surrender a short time later.

Hearing this story, it was obvious that something had gone wrong in Davis' life. The teen who had once dreamed of finishing high school and playing professional basketball ended up as another casualty in the petty and dangerous world of drugs. By all accounts he had been raised by a hard-working family, and he even had a family of his own. But the allure of easy money is often a pitfall worse than any chance encounter with a shadowy monster.

The South Carolina news media inevitably splashed Davis' photo across their front pages, highlighting the grisly details of his murder and recounting his harrowing experience with the Lizard Man. Accusations resurfaced that he may have been "high" the night he claimed to have seen the creature. But Truesdale himself, having known Davis, held fast in his belief that the young man truly saw something he thought to be a monster that night. "The Davis boy caused so much publicity that I think it affected his mind," Truesdale told us.

And who can really say. Perhaps that monster had become a demon that eventually played a part in his wayward lifestyle. The frenzy of Lizardmania was long gone, but the sensational details of Davis' encounter were never forgotten. He would always be "the kid who saw the Lizard Man." It was an infamy that almost certainly haunted him right up until the end.

Davis may be the most famous of the witnesses, but he was not the only one to meet an untimely end. In June of 1999, Johnny Blythers, the young man whose family saw a strange creature near Scape Ore Swamp in 1990, was killed in a car accident. His obituary appeared in the June 22, 1999, edition of *The State* newspaper but provided no additional details as to the circumstances. I attempted to contact the Blythers to find out more about Johnny's life after the sighting, but I had no luck tracking them down.

Around the same time, Colonel Mason Phillips died of mysterious causes. He was older than the young men who witnessed the creature, but nevertheless, his passing further established a peculiar pattern of death surrounding anyone who had seen the Lizard Man.

Next, on October 27, 2009, George Plyler passed away. By all accounts this was from natural causes, so it doesn't necessarily add to the mystery, only to the totality of deceased witnesses.

It may not be as dramatic as the legendary curse of Tutankhamen's tomb, but the apparent "Curse of the Lizard Man" is enough to give Truesdale a shiver. "I've often wondered if there's something to it," he told us. "It is unusual how some of the witnesses ended up dying the way they did."

It's certainly an interesting footnote to the story and one that I would ponder as we headed back to the place where it all got started.

Sunset in Scape Ore

At the end of our conversation, I turned off the digital recorder and scribbled a few final notes. We could have easily talked longer, but our stop at Truesdale's had already taken up more of the day than we had intended. After snapping a few photos, Cindy and I thanked him for his hospitality and said goodbye. Within a few minutes, we were on the road to the now familiar Browntown.

As we approached Scape Ore bridge, we could see waves of heat rising from the asphalt. It was mid-afternoon by now and the sun still had its grip on the Carolina countryside. I slowed the car and turned into a make-shift parking area where several tall trees mercifully blocked the sun. I rolled into their shadow and stopped the engine.

Cindy got out to look around. We had visited the infamous site on several occasions during our trip, but we wanted to hike into Scape Ore's interior one last time. I joined her as we walked to the edge of the channel where the water creeps under the bridge. Having learned so much of its history, we peered into its brackish depths wondering what kind of secrets it might truly hold.

After surveying the areas on both sides of the bridge, we walked to the back of the parking area where a small black pipe juts from the ground a few yards from the channel. A gurgling stream of water flowed from its end and spilled out onto the sandy soil. It was the artesian well that George Holloman drank from just before he allegedly saw the large, dark *thing* lurking in the trees back in 1987. I looked in the direction of the road. On the other side, I could see the crowded treeline where Holloman said the creature retreated when the car passed. I tried to imagine a monster hulking in the shadows. Granted I was standing there in daylight, not at dusk as he was, but regardless I could imagine the creepiness of such an encounter. Even in the light, the place resonated with an eerie desolation despite its proximity to the road.

Cindy noticed it, too. There was something about the area that

The bridge at Scape Ore Swamp pictured in 2012 (photo by the author)

seemed withdrawn from the rest of the world. Perhaps it was the coolness of the shadows contrasting with the heat of the August afternoon that led us to draw such a conclusion. Or maybe it was a side-effect resulting from our complete submersion into the local lore that was influencing our perception. Whatever it was, the place definitely felt strange and foreboding.

Beyond the artesian well, we located a small path that snaked off into the woods. It seemed like the perfect place to explore further, so we grabbed our gear, threw on our packs, and entered the purported lair of the Lizard Man.

As we walked along, the trees provided a lofty canopy. This made it more tolerable to hike on such a hot day, but we soon discovered the path itself was not so accommodating. As it turns out, it was not so much a hiking trail as it was a small game trail that had probably experienced some human foot traffic as people walked a short distance into the woods and turned around. As we progressed further, it became narrower as the understory began to erase its existence. Eventually, it succumbed to ground vines and thorns altogether, making progress difficult without weaving and bobbing through the brambles. At that point, we decided to make our own way, following the general route of the water. I only had a rudimentary map of the area, but as long as we stayed near the waterway, it would be easy enough to get back… or so we hoped.

As we continued on, I was reminded of Truesdale's statement that "coonhunters don't hunt in Scape Ore." Now we could see firsthand what he meant. The terrain was growing noticeably inhospitable as if resistant to our human intrusion.

Inhospitable swamps, however, were nothing new to us, so Cindy and I continued to explore undaunted into the early evening before finally settling in to do some quiet observation. Our chances of seeing some unknown creature were admittedly slim, but surely we wouldn't see anything at all if we kept up our noisy movement.

After a bit of searching, we found a comfortable spot of ground in front of a fallen tree. It was a place where we could sit somewhat concealed, yet still view the main swamp channel. We checked for snakes, then sat back on the old log to watch for any creatures that decided to venture down to the water. By this time, a few shadowy pockets had begun to settle within the trees. The birds were still chirping, but their songs were noticeably quieter.

We sat for a while, enjoying the solitude that Scape Ore provided. We could hear animals moving about in the brush. These were obviously small creatures, perhaps squirrels or rabbits. Eventually, a few frogs tested out their voices, as they prepared for their nightly chorus. Dusk in the deep woods is always enjoyable, and when combined with the possibilities of the unknown, truly exhilarating.

As I sat there in the encroaching darkness listening to unseen creatures, I began to reflect on the strange case of the Lizard Man. What was once a brief entry in a cryptozoology book or an old newspaper article was now a fascinating story that had come to life around me. Along the way I had expelled my initial impression that the subject was some sort of cartoonish creature best suited for comic books and acknowledged that perhaps the people of Bishopville had seen some unexplainable creature… or creatures. I could never be certain, of course, but I couldn't completely dismiss all of the testimony we had come across.

Not everyone may agree. The Lizard Man case is a complicated one, full of pitfalls much like Scape Ore Swamp itself. First, there were hoaxes involved. And hoaxes always reflect badly on the credibility of a case. However, given the wide variety of encounters, and the span in which they occurred, it doesn't seem that hoaxes can account for the entire phenomenon. The key sighting by Chris Davis is also cloaked in controversy. It's most certainly not a hoax but could be a case of mistaken identity. The other sightings, while often vague and brief, cannot be easily explained away as the work of a masquerading farmer. So if these truly occurred, then perhaps they help support Davis' insistence that it was a strange creature, not a man he encountered that night.

And what about the damaged vehicles? The evidence suggests common animals are to blame, but why would this happen to three different cars in the same general area? We might conclude these incidents are also hoaxes, but again, it just doesn't add up. The Waye incident occurred before the stories of the Lizard Man went public and the others long after. Perhaps they are simply coincidental, yet unrelated.

We must also consider any role that racism may have played in the case. Since many of the sightings occurred in Browntown—which was undoubtedly named for its African-American residents—it has led some to suggest that the whole thing was either derived from ethnic superstitions, or worse, devised to frighten a certain demographic. However, this doesn't seem to add up either. Some of the witnesses were black and some white. Some claimed to have seen the creature near Browntown, some claimed to have seen it near other sections of Scape Ore Swamp. Community names such as Browntown certainly remind us of the scars of racism, especially in the Deep South, but the fact that a creature was allegedly seen by the people who live in such a place does not give it any less credibility than if it had been seen running through a New Hampshire suburb. The witnesses either saw something unexplainable or they did not, regardless of their skin color.

The case is also muddled by the very concept of the creature itself. Whereas sightings of a lake monster, for example, mostly boil down to the question "Did that person see an undocumented animal or not?", the case of the Lizard Man begs us to ask a host of other questions related to the mind-boggling concept of a reptilian humanoid. The hypothetical paths that fan out from this case go in so many directions, they often end up like the path Cindy and I navigated through the swamp. They become choked with briars and brambles to the point where it's impossible to know where they lead.

If you believe in the possibility of Bigfoot, then perhaps the people of Bishopville should have named their creature Bigfoot Man because much of the phenomenon can be compared to sightings of this cryptid. If you don't believe Bigfoot is real, then we must conclude that witnesses such as the Blythers saw a person in a costume running across the road. They certainly saw something, because like Davis, it would just be too hard to fake the emotion when those individuals came to the Sheriff's Office to tell of their experience.

In the end, we might ask ourselves if these people saw anything at all. Could they all be lying? Surely not. Something scared Christopher Davis that night. There's just too much circumstantial evidence to support his story. The others as well didn't seem to have anything to gain by lying or making up such stories. They never signed autographs or sold T-shirts; they were simply telling what they honestly believed they saw. Of course this doesn't rule out mass hallucination, but again the testimonies as a whole do not seem derivative of this sole explanation.

But even if a few of these people saw a Bigfoot, a few of them saw a bear, and one of them was frightened by a farmer wearing a convincing costume, the Lizard Man of Lee County still exists. He's a product of the witness statements, the police files, the news reports, and the media sensation, all of which have formed a legend that will never die. Bishopville will never again be an anonymous little town in the quaint countryside of South Carolina. It will always be the place where people once saw the "Lizard Man." Just do an internet search for "Bishopville" or "Scape Ore Swamp" and it's quickly evident that whether or not this creature ever walks out of the bottomlands, its legacy will forever be stamped in the annals of American monsters. Bishopville is, and will always be, one of those little pockets in the civilized world where mysteries are alive and well.

Sheriff Truesdale once remarked: "People are always going to wonder about the Lizard Man." And I'm sure he's right. We have an innate need to reach into the shadows so that we may learn more about this strange world in which we inhabit, and perhaps more about ourselves in the process. We may never find all the answers we seek, but it's a quest that some of us cannot resist.

As Cindy and I sat there watching the sun go down in Scape Ore Swamp, we heard something splash through the water just out of sight. Chances are it was a common animal, but still something compelled me to look. So I raised my nightvision monocular to my eye.

It is my nature to seek the unknown.

Lyle Blackburn with Retired Sheriff Liston Truesdale

Cindy Lee and Lyle Blackburn at the edge of Scape Ore Swamp

Appendix

Lizard Man Incident Log

Spring 1986	Sighted by George Plyler in back of his house on Springvale Road.
October 1987	Sighted by construction worker George Holloman near Scape Ore bridge.
May 1988	Sighted by Frank Mitchell from his plane as it crossed the runway.
June 29, 1988	Attacked Chris Davis on Browntown Road.
July 14, 1988	Tom and Mary Waye reported car damage.
July 16, 1988	Chris Davis came to the Sheriff's Office to report his sighting.
July 1988	Columbia radio station WCOS-FM offered a $1 million dollar reward for the creature.
July 23, 1988	*People* magazine came to Bishopville.
July 24, 1988	Two teens, Rodney Nolf and Shane Stokes, reported that they saw a large, muscular creature dart across I-20 at Hwy 15. Around the same time, the police received a call about some unnatural howls heard in the same vicinity. They decided to investigate and found the hoaxed tracks on Bramlette Road.
July 25, 1988	Dan Rather from the *CBS Evening News* interviewed Truesdale.

July 29, 1988	*Good Morning America* did a live broadcast from Browntown with Truesdale.
Aug 5, 1988	The date Kenneth Orr, an airman from Florence (Shaw Air Force Base), claimed he shot the Lizard Man.
Aug 7, 1988	Reporter Loyd Dillon at *The Charlotte Observer* proposed a contest to give the Lizard Man a better name. $10 first prize.
Aug 11, 1988	Kenneth Orr was charged with unlawfully carrying an unregistered gun.
Aug 13, 1988	George Holloman gave an official statement of his sighting to Truesdale. Chris Davis signed autographs at Myrtle Square Mall.
Aug 18, 1988	Davis passed a polygraph test administered by Sumter Police Captain Earl Berry.
Aug 21, 1988	Drake Hogestyn (Roman Brady on *Days of Our Lives*) helped raise funds for the Lee County Cotton Festival and a community center.
Aug 26, 1988	Colonel Mason Phillips of the Army Corps of Engineers reported seeing a creature run across McDuffy Road in the Ashwood section of Lee County.
July 30, 1990	Sighted by the Blythers family as it crossed Browntown Road near Hickory Hill.
Sept 1990	Rumored sighting by a bank Vice President from Florence while he was deer hunting near Scape Ore Swamp.
Winter 1990	Sighted by Brian Elmore and Michelle Nunnery Elmore.

May 8, 1992	The date the Elmores gave their official statements to police.
Oct 2005	A woman in Newberry County (approximately 70-90 miles from the sightings in Lee County) reported seeing two strange creatures outside of her home.
Feb 28, 2008	Bob and Dixie Rawson reported car damage. A dead cow and coyote were found near the car.
June 20, 1999	Johnny Blythers was killed in a car accident.
June 17, 2009	Chris Davis was shot and killed at his home.
Oct 27, 2009	George Plyler passed away.

Police Reports

The following images are scans of the original police reports pertaining to several of the Lizard Man incidents.

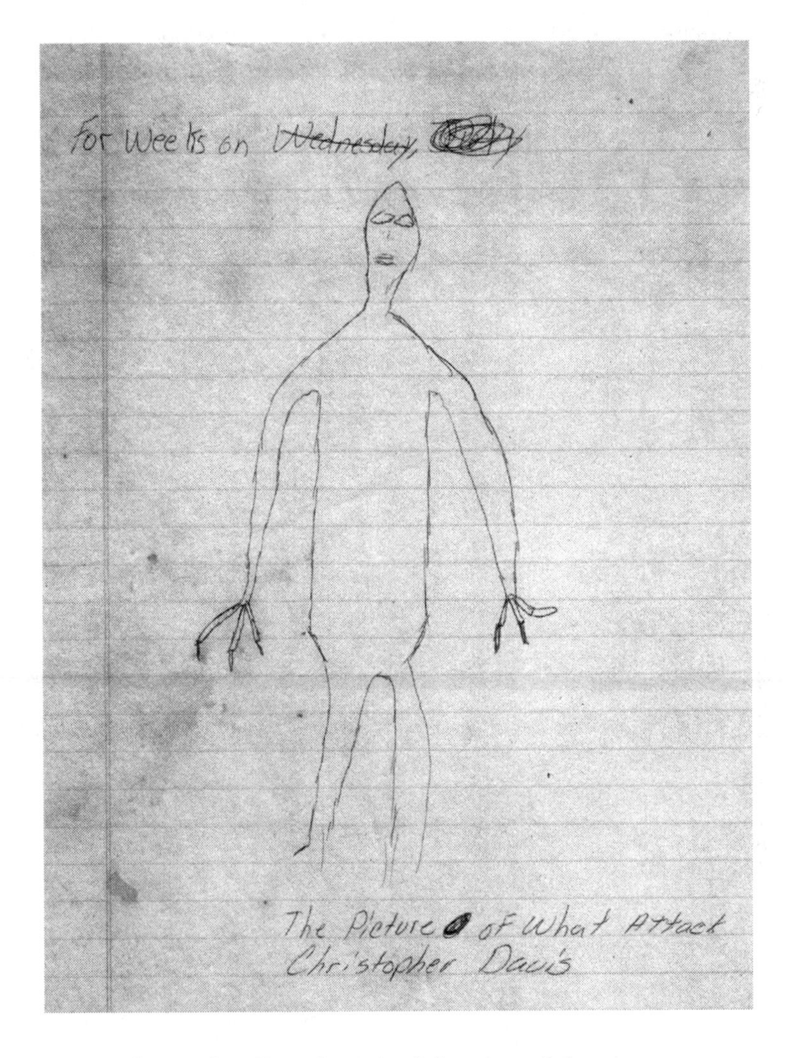

Christopher Davis' original drawing of the creature

July 31, 1990

STATEMENT OF JOHNNY BLYTHERS
ROUTE 1, BOX 163 CAMDEN, SC
D.O.B. - Oct. 17, 1971; AGE 18
JUST GRADUATED FROM BISHOPVILLE HIGH SCHOOL
EMPLOYED BY THE CITY OF BISHOPVILLE S.C.

This statement relates to an unusual sighting that happened July 30, 1990 on the Browntown-Hickory Hill Road in Lee County.

Last night about 10:30 P.M., we were coming home from the Browntown section of Lee County. It was me, my mother (Bertha Mae Blyther), two sisters, ages five and ten, my two brothers, age four (twins).

I started talking about the time we passed the flowing well in Scape Ore Swamp. I said "they aint no such thing as a Lizzard Man,.if there was, somebody would be seeing it or caught it".

We got up about a mile or mile and one half passed the butter bean shed, about 50 feet from the dirt road by those two signs, my mother was driving the car.

It was on the right side, it came out of the bushes. It jumped out in the road. My mother swerved to miss it, and mashed the brakes and sped up. It jumped out of the bushes like he was going to jump on the car. When my mother mashed the brakes, it looked like it wanted to get in the car.
SIZE: it was taller than me. I'm 5' 10". It was 6 feet tall or more.

COLOR: I would say it was brown- all of us said it was brown.

DISCRIP.: Had hair on it. Tall big eyes, brown.

I told by mother to turn around, I wanted to see what it was, but she wouldn't. We wern't going fast, about 30 miles an hour.

This was none of my imagination, I know I saw it, and I'm not trying to get on T.V.

I know " I see it last night and I ain't crazy". It stood up like a human, not straight up, but humped over, kinda in a crouched position.

Never seen anything that looked like it before. Had arms and legs-big eyes.

I ain't going to lie, I know what I see. I'm not like Chris Davis, I don't know what he saw, but I know what I see, and I'm not trying to get attention.

My mother said "Ya'll see that, ya'll see that"? My sister started screaming. We came home pretty fast. My sister thought it was going to get her.

We came home and reported it to the Sheriff's Office.

NOTE: It looked to be "what he measured to be a yard wide" says it was a lot bigger than him.

Official statement of Johnny Blythers (Part 1)

Ore Swamp. I said "they aint no such thing as a Lizzard Man, if there was, somebody would be seeing it or caught it".

We got up about a mile or mile and one half passed the butter bean shed, about 50 feet from the dirt road by those two signs, my mother was driving the car.

It was on the right side, it came out of the bushes. It jumped out in the road. My mother swerved to miss it, and mashed the brakes and sped up. It jumped out of the bushes like he was going to jump on the car. When my mother mashed the brakes, it looked like it wanted to get in the car.
SIZE: it was taller than me. I'm 5' 10". It was 6 feet tall or more.

COLOR: I would say it was brown- all of us said it was brown.

DISCRIP.: Had hair on it. Tall big eyes, brown.

I told by mother to turn around, I wanted to see what it was, but she wouldn't. We wern't going fast, about 30 miles an hour.

This was none of my imagination, I know I saw it, and I'm not trying to get on T.V.

I know " I see it last night and I ain't crazy". It stood up like a human, not straight up, but humped over, kinda in a crouched position.

Never seen anything that looked like it before. Had arms and legs-big eyes.

I ain't going to lie, I know what I see. I'm not like Chris Davis, I don't know what he saw, but I know what I see, and I'm not trying to get attention.

My mother said "Ya'll see that, ya'll see that"? My sister started screaming. We came home pretty fast. My sister thought it was going to get her.

We came home and reported it to the Sheriff's Office.

NOTE: It looked to be "what he measured to be a yard wide" says it was a lot bigger than him.

The people I talked to , they worked with me, (some of them)-that I talked to - today were laughing at me. They said "it was my imagination".

I saw it before she (my mother) mashed the brakes and I saw it after. I think if she hadn't swerved, we would have hit it.

I certainly don't think it was a human dressed in something, because no human ain't going to jump in front of a car.

It was definately not a bear.

This ends the statement of Johnny Blythers.

WITNESS
July 31, 1990

Official statement of Johnny Blythers (Part 2)

August 2, 1990

STATEMENT OF BERTHA MAE BLYTHERS
ROUTE 1, BOX 163, CAMDEN, S.C.
D.O.B. - JAN. 19, 1954; AGE 36
EMPLOYED AT BURLINGTON PLANT IN BISHOPVILLE, S.C.
HOUSE WIFE, SEPERATED - FOUR CHILDREN.

This statement relates to an unusual sighting that took place July 30, 1990 on the Browntown-Hickory Hill Road of Lee County.

This past Monday night I went to my mother's house in Browntown to pick up my son. We went to McDonald's on Highway 15 near Bishopville to get something to eat. We left there about 20 minutes after 10:00 P.M., was headed home and came through Browntown and Scape Ore Swamp.

We were laughing, singing and eating french fries and listening to the radio, and started talking about the Lizzard Man, right be fore we got to that bridge. After we got through talking about him, we were laughing and singing again, one big happy family.

The people in the car was my son, Johnny, age 18, my daughters, Tamacia, age 11, Christa, age 5, and my two twin boys, age 4.

We passed the bridge and was down the road near a mile. I was looking straight ahead going about 25 M.P.H., and I saw this big brown thing, it jumped up at the window. I quickly sped up and went on the other side of the road to keep him from dragging my 11 year old girl out of the car. I didn't see with my lights directly on it. It nearly scared me to death.

After I turned to the other side of the road, I mashed my brakes and my son saw something walking across the road behind us. My son wanted to turn around and go back, but I wouldn't. "I was too scared. There was no way I would have turned that car around and went back". I came on home and called my sister, Virginia, and told her about it. She told me to report to the Sheriff's Office.

I didn't want to report it, because they would think it was a joke or something. But I told them whether they believed me or not.

My son wanted me to turn around right then and report it, but I came on home, I was scared to get out of the car, and when I did I liked to never got the right key to get in the door.

<u>DESCRIPTION</u>: It was tall, had a chest like a human and two arms like a human. I saw from the waist up, but I didn's see his face, the chest was big - NOTE: (SHE MEASURED A YARD WIDE) it almost filled the window on the passenger's side. NOTE: (SHE WAS DRIVING A 1980 BUICK REGAL) What I saw of it, looked like hair. NOTE: It was about the color brown as my car. NOTE: (which is between gold and brown).
I'm not wanting to get on T.V. It was not my imagination, I know I seen it and it wasn's a deer. I never seen anything like it before. It wasn't a deer or a bear my 11 year old daughter was screaming so, I had to reach over and straighten her out. It definately was not a person either.

This ends the statement of Bertha Mae Blythers.

WITNESS [signature]
August 2, 1990

Official statement of Bertha Mae Blythers

Page # 1

August 2, 1990 STATEMENT OF TAMACIA BLYTHERS AGE 11
Route # 1, Box 163 Camden, S.C. DOB 3018079
Lives with her mother and goes to school

We were riding along. I was the first to see it. I was on the the door on the passengers side. I had the glass down. I had my arm on the door where the glass was at. When I saw it, I hollowerd and ran the glass up, and got close to my mother.

DESCRIPTION: Tall-Taller than the car, Brown looking, a big chest had big eyes...had two arms, Don't know how his face looked...first seen his eyes (Never seen nothing like it before. I didn't see a tail)

Mother says if she hadn't whiped over he would have hit her car or jumped on it. Would like to see it and get a picture.

Mother said she was so scared her body light and she held her heart all the way home.

First time any of them has ever seen anything at night... Careful about dogs not to run over them.

This ends the statement of Tamacia Blythers.

Witness: Liston Truesdale

August 2, 1990

Official statement of Tamacia Blythers

VOLUNTARY STATEMENT

DATE August 12 1988 PLACE Sheriffs office TIME STARTED 11:04 A.M.

I, the undersigned, Kenneth Bruce Orr, am 26 years of age, my date and place of birth being the 12 day of April 1962 at Newnan Ga.

I now live at Rt 4, Box 493 Florence, S.C.

Before answering any questions or making any statements, Liston Truesdale and Jim Clardy

a person who identified himself as a Sheriff and Deputy Sheriff

duly warned and advised me, and I know and understand that I have the following rights: That I have the right to remain silent and I do not have to answer any questions or make any statements at all; that any statement I make can and will be used against me in a court or courts of law for the offense or offenses concerning which the following statement is hereinafter made; that I have the right to consult with a lawyer of my own choice before or at anytime during any questioning or statements I make; that if I cannot afford to hire a lawyer, I may request and have a lawyer appointed for me by the proper authority, before or at anytime during any questioning or statements that I make, without cost or expense to me; that I can stop answering any questions or making any statements at any time that I choose, and call for the presence of a lawyer to advise me before continuing any more questioning or making any more statements, whether or not I have already answered some questions or made some statements.

I do not want to talk to a lawyer, and I hereby knowingly and purposely waive my right to remain silent, and my right to have a lawyer present while I make the following statement to the aforesaid person, knowing that I have the right and privilege to terminate any interview at any time hereafter and have a lawyer present with me before answering any more questions or making any more statements, if I choose to do so.

I declare that the following voluntary statement is made of my own free will without promise of hope or reward, without fear or threat of physical harm, without coercion, favor or offer of favor, without leniency or offer of leniency, by any person or persons whomsoever.

After all of these rights were explained to me I want to tell the truth what happened. nothing actually happened that day. on August 5th 1988. I made the Report just to keep the Legend of the Lizzard man alive.
I took the pistol out of the glove box to show to the deputy that came to the truck stop. He stated he didn't need to see it so I laid it on the seat. When I got back to the car I put it back in the glove box and came to the Sheriffs office. the only other time I took it out the box was to give it the deputy Sheriff I did not fire that weapon that morning

I have read each page of this statement consisting of 2 page(s), each page of which bears my signature, and corrections, if any, bear my initials, and I certify that the facts contained herein are true and correct. I further certify that I made no request for the advice or presence of a lawyer before or during any part of this statement, nor at any time before it was finished did I request that this statement be stopped. I also declare that I was not told or prompted what to say in this statement.

This statement was completed at 12:15 PM on the 12th day of August, 1988.

WITNESS: Liston Truesdale

Kenneth R Orr

Voluntary statement of Kenneth Orr

8-9-90 Re: Lizzard Man Call—

Statement of Ed Corey - Deputy Sheriff.

I went to the Jail - and Shirley Bryant's gave me the complaint card - where Bertha Mae Blythers & family saw something 6 feet tall, Brown

I went down there & check every thing on the Road near the butler bean shed — couldn't find anything - had Shirley to call her back to have her meet me at the butler bean shed

She met me out there. the other end of [illegible] Road - and she showed me where it was at or where they saw it.

She said "she was going about 30 mph & it was standing along side of the Road and it Reached out & looked like it was going to Reach in & get her Baby—

I checked and could not find any tracks in the field, but the edge of the Road was vines and grass - and couldn't see anything where the grass was.

She was nervous & excited - and there was no doubt - she saw something standing there — all I did was let her show me and she was Really wanting to get out there—

Deputy Ed Corey's notes regarding his investigation of the Blythers sighting

Swamp Monster Movie List

For those of you who wish to delve into the wonderful world of swamp monster cinema, I have assembled a list of movies that feature reptilian or amphibious-type creatures. As you will see, this is not confined to the category of horror, but I did draw the line when it came to movies set in outer space.

Creature from the Black Lagoon (1954)
Revenge of the Creature (1955)
The Creature Walks Among Us (1956)
The Alligator People (1959)
The Monster of Piedras Blancas (1959)
The Hideous Sun Demon (1959)
The Horror of Beach Party (1964)
Curse of the Swamp Creature (1966)
The Reptile (1966)
Graveyard of Horror (1971)
Zaat (1971)
Night of the Cobra Woman (1972)
Sssssss (1973)
Track of the Moonbeast (1976)
When the Screaming Stops a.k.a. *The Loreley's Grasp* (1976)
Screamers (1979)
Humanoids from the Deep (1980)
Conan the Barbarian (1982)
Bog (1983)
The Legend of Gator Face (1996)
Lizard Baby (2004)
Land of the Lost (2009)
Creature (2011)
The Amazing Spider-Man (2012)

Acknowledgments

Thanks to the following people for their invaluable assistance and contributions to this book:

Dave Alexander, Chad Arment, Gee Atkinson, Sandy Blackburn, Rachael Bradbury, Chris Buntenbah, Carlos Cabrera, Dave Coleman, Loren Coleman, John "Wolf" Covington, Janson Cox, Loyd Dillon, Josh Gates, Ken Gerhard, Micah Hanks, Jerry Hestand, Patrick Huyghe & Anomalist Books, Jessica James, Chase Kloetzke, Cindy Lee, Marvin Leeper, Alicia Lutz, Jesse Martin, Dr. Jeff Meldrum, Frank Mitchell, Bart Nunnelly, Joseph Patterson, Charlie Raymond, Nick Redfern, Mike Richburg, Skeet Woodham, John Rhodes, Rob Robinson, Jeff Sisson, Lon Strickler, David Weatherly, Craig Woolheater, South Carolina Cotton Museum, and the family at Harry & Harry Too restaurant.

Special thanks to Liston Truesdale for his hospitality, insight, and access to historical documents.

Bibliography

Books

Coleman, Loren. *Mothman and Other Curious Encounters*. New York. Paraview Press, 2002. pp. 88-97.

Coleman, Loren. *Mysterious America*. New York. Paraview Pocket Books. 2007. pp. 280-282.

Coleman, Loren and Patrick Huyghe. *The Field Guide to Bigfoot and Other Mystery Primates*. San Antonio. Anomalist Books, 2006. pp. 62-64.

Gregoire, Anna King. *History of Sumter County, South Carolina*. Sumter: Library Board of Sumter County, 1954.

Glut, Donald F. *Classic Movie Monsters*. University of Michigan. Scarecrow Press, 1978.

Hall, Mark A. *Lizardmen: The True Story of Mermen and Mermaids*. North Carolina. Mark A. Hall Publications, 2005. pp. 6, 33-34, 49, 53.

Icke, David. *The Biggest Secret*. David Icke Books. 1999. pp. 19-25.

Keel, John A., *Strange Creatures From Time and Space*. Connecticut. Fawcett Publications, Inc, 1970. pp. 93, 101, 107.

Mooney, James. *The Sacred Formulas of the Cherokees*. Seventh Annual Report of the Bureau of Ethnology to the Secretary of the Smithsonian Institution, 1885-1886. Government Printing Office, Washington. 1891. pp. 301-398.

Swanton, John R., *Early History of the Creek Indians and Their Neighbors*. Pub. Smithsonian Institution, Bureau of American Ethnology, Bulletin 73. Washington, 1922.

Historical Documents

Sumter County Conveyances. Book A. March 1803.

Journals

Neuffer, Claude Henry, ed. *Names in South Carolina.* University of South Carolina Department of English. Vol. 12 No. 10. Winter 1965.

Russell, D. A.; Séguin, R., (1982). "Reconstruction of the Small Cretaceous Theropod Stenonychosaurus inequalis and a Hypothetical Dinosauroid." Syllogeus No. 37: 1–43.

Magazines

Lutz, Alicia. "Off To See the Lizard," *College of Charleston Magazine* (September 2011).

Staff writer. "The Legend of Lizard Man." *Time* (August 1988).

Weaver, Tom. "Producer from the Black Lagoon," *Starlog* no. 218 (September 1995).

Newspapers

Associated Press. "He's baaack: Lee sheriff reports Lizard Man sighting." *State* 02 Sept. 1988.

Associated Press. "Lizard Man Sightings Spur Hunt." *Oxnard Press Courier* 20 July 1988.

Associated Press. "Lizard man turns Lee County green from sales spree." *Item* 24 July 1988.

Atkinson, Gee. "Suspicion shifts from swamp to kennel." *Lee County Observer* 21 May 2008.

Avery, Michael V., "Gail Gaddy braves the unknown at Scape Ore Swamp." *News Herald* [Morganton, NC] 11 Aug. 1988.

Burns, Randy. "Attack on Lee County couple's van resembles 1988 incident." *Item* 04 Mar. 2008.

Burns, Randy. "DNA debunks Lizard Man attack." *Item* 15 May 2008.

Burns, Randy. "Lizard Man replaces Cotton Festival." *Item* 2008.

Burns, Randy. "Police: Murder victim Lizard Man witness." *Item* 20 June 2009.

Burns, Randy. "Tall tales of a man with scales." *Item* 21 July 2004.
Burns, Randy. "TV crew shooting show on Lizard Man." *Item* 07 Dec. 2004.
Cheek, Larry. "Lizard Man: A Report." *Observer-Times* 20 Sept. 1990.
Columnist. "Lizard man stories tell of dangerous creature." *Augusta Chronicle* 19 Nov. 2000.
Dillon, Loyd. "Is Lizard Man A Croc?" *Charlotte Observer* 07 Aug. 1988.
Georgas, George. "Bumps in the night? Lee County seeks 'lizard' with bad attitude." *Item* 20 July 1988.
Georgas, George. "Lee's Lizard man left his mark." *Item* 01 Jan. 1989.
Georgas, George. "The five sightings." *Item* 01 Aug. 1988.
Georgas, George. "The saga continues: Lee lizard tale grows." *Item* July 1988
Giddens, Tharon A., "Lee County residents profit from mysterious sightings." *Anderson Independent-Mail* (date unknown).
Greene, Lisa. "It's a Big, Green, Money-Making Machine." *Charlotte Observer* 14 Aug. 1988.
Lewis, Michael. "Lizard Man: S.C.'s answer to Bigfoot." *State* 15 Aug. 1988.
McGirt, Tonyia. "Television crew tracks Bishopville's Lizard Man. *Item* 07 Apr. 1989.
McLeod, Lisa S., "Large Tracks Found in Browntown." *Lee County Observer* July 1988.
McLeod, Lisa S., "Lizard Business Booming In County." *Lee County Observer* 03 Aug. 1988.
McLeod, Lisa S., "Sheriff Tired Of Lizard Man, But Admits Good For County." *Lee County Observer* 17 Aug. 1988.
McLeod, Lisa S., "Unidentified 'High Official' Reports Lizard Man Sighting. *Lee County Observer* 31 Aug. 1988.
McLeod, Lisa S. and Roger McKenzie. "The Lizard Man Stalks Browntown Section of County." *Lee County Observer* 20 July 1988.
Monk, John. "Fame Follows Close Encounter Of The Lizard Kind." *Charlotte Observer* 02 Aug. 1988.
Moniz, Dave. "Lizard Man is gone and mostly forgotten." *State* 08 Apr. 1988.
Rhyne, Debbie. "Lizard Man still grips imaginations." *Item* 31 Oct. 1998.
Schafer, Matt. "Legends, lore and the Lizard Man." *Item* 27 June 2001.

Smith, Bruce. "The Lizard Man: Rumors of revolting reptile run rampant." *News Herald* [Morganton, NC] 09 Aug. 1988.
Staff writer. "Lizard Man shooter charged with unlawfully carrying gun." *Item* 12 Aug. 1988.
Staff writer. "To keep a monstrous legend alive." *Houston Chronicle* 13 Aug. 1988.
Staff writer. "Soap actor to probe Lizard Man mystery." *State* 19 Aug. 1988.
Staff writer. "Lee County Cotton Festival." *Item* 22 Aug. 1988.
Staff writer. "A Dangerous Nuisance." *Lee County Observer* 20 July 1988.
Staff writer. "Lizard Man has this town eyeing swamp." *News-Journal* [Florence, SC] 21 July 1988.
Staff writer. "Lizard mania grabbed Lee County by the tail. *Item* 16 July 1988.
Staff writer. "'Haunted' Ax Found, But Legend Lives On." *Lee County Observer* (date unknown).
Staff writer. "Swamp filled with curious onlookers, bounty hunters." *Anderson Independent-Mail* 31 July 1988.
Scales, Walter Ed. "Does a Monster Lurk in Lake Conway Waters?" *Log Cabin Democrat* 7 Mar. 1952.
Strong, Willard. "Lizard Man's Legacy." *News and Courier* [Charleston, SC] 21 July 1988.
Tuten, Jan. "Lizard Man tale is a real sob story." *State* 23 Oct. 1988.
Tuten, Jan. "Big tracks discovered in 'Lizard Man' area." *Spartanburg Herald Journal* 26 July 1988.
Tuten, Jan. "Florence man says he wounded 'Lizard Man'." *State* 06 Aug. 1988.
Tuten, Jan. "Man who shot 'Lizard Man' faces gun carrying charges." *State* 12 Aug. 1988.
Tuten, Jan. "Lizardless summer." *State* 24 July 1989.
Tuten, Jan. "Lizard Man peacock theory making rounds in Lee County." *Spartanburg Herald-Journal* 25 July 1991.
Tuten, Jan. "Even some Carolina cops spooked by 'Lizard Man'." *Knight-Ridder Newspapers* (date unknown).
United Press International. "7-foot 'lizard man' attacks cars." *Modesto Bee* [Modesto, CA] 20 July 1988.
United Press International. "Lizard Man Reported Prowling Swamps." *Korea Times* (date unknown).

United Press International. "S.C. man admits 'Lizard Man' shooting story a hoax." *Telegraph* [North Platte, NE] 13 Aug. 1988.
Wecker, David. "True Tale of Loveland's Monster." *Cincinnati Post* 14 Aug. 1999.

Online Articles

Gross, Patrick. "June or July 1955, Loveland, Ohio, USA, Carlos Flannigan." *URECAT - UFO Related Entities Catalog*. 14 Feb. 2008. Web.
Hendricks, Richard. "Highway 13's Reptile Man." *Weird Wisconsin*. 06 Sept. 2001. Web.
Larson, Ron. "Types of Swamps." *Arbor Day Foundation*. (Not dated.) Web.
Martin, Jesse. "The Thetis Lake Monster—A First-hand Encounter." *World Fishing Network*. 01 Nov. 2011. Web.
Mast, M.A., and Turk, J.T., "Scape Ore Swamp near Bishopville, South Carolina (Station 02135300)." *USGS: science for a changing world*. 31 July 2000. Web.
Naish, Darren. "Dinosauroids Revisited." *Darren Naish: Tetrapod Zoology*. 02 Nov. 2006. Web.
Talarico, Lauren. "Lizard man Returns to Lee County?" *WLTX 19 News*. 06 July 2011. Web.
Tordjman, Dan (2008). "Dead cow, coyote found near site of "Lizard Man" mystery." *WISTV 10 News*. 28 Apr. 2008. Web.
Stafford, Jeff. "Creature from the Black Lagoon." *Turner Classic Movies: Film Article*. (Not dated.) Web.
"Gatormen: Florida-Louisiana, USA." *American Monsters*. (Not dated.) Web.
"List of reptilian humanoids." *Wikipedia: The Free Encyclopedia*. Wikimedia Foundation, Inc. 28 Feb. 2013. Web.
"Thetis Lake Theories." CFZ Canada. *Centre for Fortean Zoology*. 18 July 2012. Web.

Press Releases

Beckjord, Eric. "Brief Review of Incidents Relating to the South Carolina Bigfoot Case 1988." The Cryptozoology Museum Project Press Release - July 27, 1988.

Websites

American Monsters: http://www.americanmonsters.com
Bigfoot Field Researchers Organization: http://www.bfro.net
Circus Historical Society: http://www.circushistory.org
Internet Movie Database, The: http://www.imdb.com
Reptoids Research Center: http://www.reptoids.com
South Carolina's Information Highway: http://www.sciway.net
Vanderbilt Television News Archive: http://tvnews.vanderbilt.edu

Index

Italic page numbers refer to illustrations

About the Author

Lyle Blackburn is an author, musician, and cryptid researcher from Texas. He has always been fascinated with legends, lore, and sighting reports of "real-life monsters," and is the author of the *The Beast of Boggy Creek: The True Story of the Fouke Monster*. During his research, Lyle has often explored the remote reaches of the southern U.S. in search of shadowy creatures said to inhabit the dense backwoods and swamplands of these areas.

Lyle is also a featured speaker at cryptozoology and horror conferences around North America. He has been heard on numerous radio programs, including *Coast To Coast AM*, and appeared on television shows such as *Monsters and Mysteries in America* and the *CBS Sunday Morning Show*.

For more information, visit: www.lyleblackburn.com

CPSIA information can be obtained at www.ICGtesting.com
Printed in the USA
BVOW02s1754281214

380957BV00009B/262/P

9 781938 398162